PID Passivity-Based Control of Nonlinear Systems with Applications

IEEE Press
445 Hoes Lane
Piscataway, NJ 08854

PID Passivity-Based Control of Nonlinear Systems with Applications

Romeo Ortega
Instituto Tecnológico Autónomo de México

José Guadalupe Romero
Instituto Tecnológico Autónomo de México

Pablo Borja
University of Groningen

Alejandro Donaire
The University of Newcastle

IEEE PRESS
WILEY

Published by John Wiley & Sons, Inc., Hoboken, New Jersey.
Published simultaneously in Canada.

Library of Congress Cataloging-in-Publication Data

Names: Ortega, Romeo, 1954- author. | Romero, J. G., author. | Borja, Pablo, author. | Donaire, Alejandro, author.
Title: PID passivity-based control of nonlinear systems with applications / Romeo Ortega, José Guadalupe Romero, Pablo Borja, Alejandro Donaire.
Description: Hoboken, New Jersey : Wiley-IEEE Press, [2021] | Includes index.
Identifiers: LCCN 2021024088 (print) | LCCN 2021024089 (ebook) | ISBN 9781119694168 (cloth) | ISBN 9781119694175 (adobe pdf) | ISBN 9781119694182 (epub)
Subjects: LCSH: PID controllers. | Nonlinear systems.
Classification: LCC TJ223.P55 O66 2021 (print) | LCC TJ223.P55 (ebook) | DDC 629.8/36–dc23
LC record available at https://lccn.loc.gov/2021024088
LC ebook record available at https://lccn.loc.gov/2021024089

Cover Design: Wiley
Cover Image: © marty8801/iStock/Getty Images

Set in 9.5/12.5pt STIXTwoText by Straive, Chennai, India

10 9 8 7 6 5 4 3 2 1

To the memory of my mother – Romeo.

To the memory of my mother; and to Salua, Patricia, Liliana and Vianca with all my love – José Guadalupe.

To Celia with all my love. To my parents, Diana and Mario – Pablo.

To Delfina, Phoebe and Cristina – Alejandro.

Contents

Author Biographies

Romeo Ortega was born in Mexico. He obtained his BSc in Electrical and Mechanical Engineering from the National University of Mexico, Master of Engineering from Polytechnical Institute of Leningrad, USSR, and the Docteur D'Etat from the Polytechnical Institute of Grenoble, France in 1974, 1978 and 1984 respectively.

He then joined the National University of Mexico, where he worked until 1989. He was a Visiting Professor at the University of Illinois in 1987–1988 and at McGill University in 1991–1992, and a Fellow of the Japan Society for Promotion of Science in 1990–1991. He was a member of the French National Research Council (CNRS) from June 1992 to July 2020, where he was a "Directeur de Recherche" in the Laboratoire de Signaux et Systemes (CentraleSupelec) in Gif-sur-Yvette, France. Currently, he is a full time Professor at ITAM in Mexico. His research interests are in the fields of nonlinear and adaptive control, with special emphasis on applications.

He has published five books and more than 350 scientific papers in international journals, with an h-index of 84. He has supervised more than 35 PhD thesis. He is a Fellow Member of the IEEE since 1999 (Life 2020) and an IFAC Fellow since 2016. He has served as chairman in several IFAC and IEEE committees and participated in various editorial boards of international journals. He is currently Editor in Chief of International Journal of Adaptive Control and Signal Processing and Senior Editor of Asian Journal of Control.

José Guadalupe Romero was born in Tlaxcala, Mexico in 1983. He received the BS degree in electronic engineering from the University of Zacatecas, Zacatecas, Mexico, in 2006 and the MSc degree in robotics and advanced manufacturing from the Centre for Research and Advanced Studies, National Polytechnic Institute (CINVESTAV), Mexico, in 2009. He obtained the PhD degree in Control Theory from the University of Paris-Sud XI, France in 2013.

He was a postdoctoral fellow at Schneider electric and EECI in Paris France and at the Laboratoire d'Informatique, de Robotique et de Microélectronique de Montpellier (LIRMM) in 2014 and 2015, respectively. Currently, he is an associate professor and researcher at the Digital Systems Department at the Instituto Tecnológico Autónomo de México (ITAM) and since 2019 he is the Director of undergraduate mechatronics engineering program.

His research interests are focused on nonlinear and adaptive control, stability analysis and the state estimation problem, with application to mechanical systems, aerial vehicles, mobile robots and multi-agent systems.

Pablo Borja was born in Mexico City, Mexico. He obtained the B.Eng. in Electrical and Electronic Engineering and the M.Eng. from the National Autonomous University of Mexico in 2011 and 2014, respectively, and the PhD in Control Systems from the University Paris Saclay, France, in 2017.

From 2017 to 2018, he was a postdoctoral researcher member of the Engineering and Technology Institute Groningen (ENTEG) at the University of Groningen (RUG). Since 2018 he is a fellow of the Faculty of Science and Engineering and ENTEG member at the RUG. His research interests encompass the control and analysis of nonlinear systems, passivity-based control, control of physical systems, passivity and its role in control theory, and model reduction.

Alejandro Donaire received the Electronic Engineering and PhD degrees in 2003 and 2009, respectively, from the National University of Rosario, Argentina. His work was supported by the Argentine National Council of Scientific and Technical Research, CONICET. In 2009, he joined the Centre for Complex Dynamic Systems and Control at The University of Newcastle, Australia, and in 2011, he received the Postdoctoral Research Fellowship of the University of Newcastle, Australia. From 2015 to March 2017 he was with the PRISMA Lab at the University of Naples Federico II, and from 2017 to 2019 with the Institute for Future Environments, School of Electrical Engineering and Computer Science, Queensland University of Technology, Australia. In 2019, he joined the School of Engineering, The University of Newcastle, Australia, where he conducts his academic activities. His research interests include nonlinear and energy-based control theory with application to electrical drives, multi-agent systems, robotics, smart micro-grids networks, marine and aerospace mechatronics, and power systems.

Preface

It is widely recognized that proportional-integral-derivative (PID) control offers the simplest and yet most efficient solution to many real-world control problems. It is said to be a *universal* controller in the sense that the integral action takes care of the past, the proportional one of the present, and the derivative term has a predictive effect. Since the invention of PID control in 1910, the popularity of PID control has grown tremendously (Ang et al., 2005; Åstrom and Hägglund, 1995, 2006; Samad, 2017).

It is interesting to quote a 2018 report of Karl Åstrom (Åstrom, 2018) where he points out the following:

- In spite of the predictions that other control techniques, e.g. model predictive control (MPC), will make, the PID obsolete, more than 90% of industrial controllers are still implemented based around PID algorithms.
- In a report of Bill Bialkowski of the Canadian consulting company Entech, it is indicated that out of 3000–5000 control loops in the paper mill industry, 97% use proportional-derivative (PI) and the remaining 3% are MPC, adaptive, etc.
- In the same report it is indicated that, out of the 50% of the PIDs that *do not work* well, 30% are due to *bad tuning*.

Indeed, it is very well known that PID controllers yield, in general, a satisfactory performance provided they are *well tuned*. The need to fulfill this requirement has been the major driving force of the research on PID control, with the vast majority of the reports related to the development of various PID tuning techniques, which are customarily based on a *linear approximation* of the plant around a fixed operating point (or a given trajectory). When the range of operation of the system is large, the linear approximation is invalidated and the procedure to tune the gains of PID regulators is a challenging task. Although gain scheduling, auto tuning, and adaptation provide some help to overcome this problem, they suffer from well-documented drawbacks that include being time-consuming and fragility of the design. The interested reader is referred to Ang et al. (2005) for a recent, detailed account of the various trends and topics pertaining to PID tuning.

The present book is devoted to the study of PID *passivity-based control* (PBC), which provides a solution to the tuning problem of PID control of *nonlinear systems*. The underlying principle for the operation of PID-PBC is, as the name indicates, the property of *passivity*, which is a fundamental property of dynamical systems. One of the foundational results of control theory is the *passivity theorem* (Desoer and Vidyasagar, 2009; Khalil, 2002; van der Schaft, 2016), which states that the feedback interconnection of two passive systems ensures convergence of the output to zero and stability (in the $\mathcal{L}_2$ sense) of the closed-loop. On the other hand, it is well-known (van der Schaft, 2016) that PID controllers are (output strictly) passive systems – *for all* positive PID gains. Therefore, wrapping the PID around a passive output yields a stable system for all PID gains. Clearly, this situation simplifies the gain-tuning task, since the designer is left with the only task of selecting, among all positive gains, those that ensure the best *transient performance*.

PID-PBCs have been successfully applied to a wide class of physical systems, see e.g. Aranovskiy et al. (2016), Castaños et al. (2009), Cisneros et al. (2015, 2016), De Persis and Monshizadeh (2017), Hernández-Gómez et al. (2010), Meza et al. (2012), Romero et al. (2018), Sanders and Verghese (1992), and Talj et al. (2010). However, their application has mainly been restricted to academic circles. It is the authors' belief that PID-PBCs have an enormous potential in engineering practice and should be promoted among practitioners. The main objective of the book is then to give prospective designers of PID-PBCs the tools to successfully use this technique in their practical applications. Toward this end, we provide a basic introduction to the theoretical foundations of the topic, keeping the mathematical level at the strict minimum necessary to cover the material in a rigorous way, but at the same time to make it accessible to an audience more interested in its practical application. To fulfill this objective, we have skipped technically involved theoretical proofs – referring the interested reader to their adequate source – and we have included a large number of practical examples.

We are aware that aiming at penetrating current engineering practice is a very challenging task. It is our strong belief that combining the unquestionable dominance of PIDs in applications with the fundamental property of passivity, which in the case of physical systems captures the universal feature of energy conservation, yields an unbeatable argument to justify its application.

Bibliography

K. H. Ang, G. Chong, and Y. Li. PID control system analysis, design and technology. *IEEE Transactions on Control Systems Technology*, 13(4): 559–576, 2005.

S. Aranovskiy, R. Ortega, and R. Cisneros. A robust PI passivity-based control of nonlinear systems and its application to temperature regulation. *International Journal of Robust and Nonlinear Control*, 26(10): 2216–2231, 2016.

K. J. Åstrom. Advances in PID control. In *XXXIX Jornadas de Automatica*, Badajoz, Spain, 2018.

K. J. Åstrom and T. Hägglund. *PID Controllers: Theory, Design, and Tuning*. 2nd edition. Instrument Society of America, 1995.

K. J. Åstrom and T. Hägglund. *Advanced PID control*. ISA-The Instrumentation, Systems, and Automation Society, Research Triangle Park, NC 27709, 2006.

F. Castaños, B. Jayawardhana, R. Ortega, and E. García-Canseco. Proportional plus integral control for set point regulation of a class of nonlinear RLC circuits. *Circuits, Systems and Signal Processing*, 28(4): 609–623, 2009.

R. Cisneros, M. Pirro, G. Bergna-Díaz, R. Ortega, G. Ippoliti, and M. Molinas. Global tracking passivity-based PI control of bilinear systems and its application to the boost and modular multilevel converters. *Control Engineering Practice*, 43(10): 109–119, 2015.

R. Cisneros, R. Gao, R. Ortega, and I. Husain. PI passivity-based control for maximum power extraction of a wind energy system with guaranteed stability properties. *International Journal of Emerging Electric Power Systems*, 17(5): 567–573, 2016.

C. De Persis and N. Monshizadeh. Bregman storage functions for microgrid control. *IEEE Transactions on Automatic Control*, 63(1): 53–68, 2017.

C. A. Desoer and M. Vidyasagar. *Feedback Systems: Input-Output Properties*. Academic Press, New York, 2009.

M. Hernández-Gómez, R. Ortega, F. Lamnabhi-Lagarrigue, and G. Escobar. Adaptive PI stabilization of switched power converters. *IEEE Transactions on Control Systems Technology*, 18(3): 688–698, 2010.

H. Khalil. *Nonlinear Systems*. Prentice-Hall, Upper Saddle River, NJ, 2002.

C. Meza, D. Biel, D. Jeltsema, and J. M. A. Scherpen. Lyapunov-based control scheme for single-phase grid-connected PV central inverters. *IEEE Transactions on Control Systems Technology*, 20(2): 520–529, 2012.

J. G. Romero, A. Donaire, R. Ortega, and P. Borja. Global stabilisation of underactuated mechanical systems via PID passivity-based control. *Automatica*, 96(10): 178–185, 2018.

T. Samad. A survey on industry impact and challenges thereof. *IEEE Control Systems Magazine*, 37(1): 17–18, 2017.

S. R. Sanders and G. C. Verghese. Lyapunov-based control for switched power converters. *IEEE Transactions on Power Electronics*, 7(1): 17–24, 1992.

R. Talj, D. Hissel, R. Ortega, M. Becherif, and M. Hilairet. Experimental validation of a PEM fuel cell reduced order model and a moto-compressor higher order sliding mode control. *IEEE Transactions on Industrial Electronics*, 57(6): 1906–1913, 2010.

A. J. van der Schaft. *L_2-Gain and Passivity Techniques in Nonlinear Control.* Springer-Verlag, Berlin, 3rd edition, 2016.

Acknowledgments

This book is the result of extensive research collaborations during the last 10 years. Some of the results of these collaborations have been reported in the papers (Bergna-Díaz et al., 2019; Borja et al., 2016, 2020; Castaños et al., 2009; Chang et al., 2000; Cisneros et al., 2013, 2015, 2016, 2020; Donaire and Junco, 2009; Donaire et al., 2016a, 2017; Escobar et al., 1999; Ferguson et al., 2017a; Ferguson et al., 2017b, 2018, 2020; Gandhi et al., 2016; Hernández-Gómez et al., 2012; Jaafar et al., 2013; Jayawardhana et al., 2007; Jung et al., 2015; Monshizadeh et al., 2019; Ortega et al., 2020; Pérez et al., 2004; Talj et al., 2009, 2010, 2011; Wu et al., 2020; Zhang et al., 2015, 2018; Zonetti and Ortega, 2015; Zonetti et al., 2015). We are grateful to our co-authors, S. Aranovskiy, A. Astolfi, A. Allawieh, D. Bazylev, M. Becherif, A. Benchaib, G. Bergna-Díaz, A. Bobtsov, F. Castaños, G. Chang, R. Cisneros, M. Crespo, G. Duan, D. Efimov, G. Escobar, G. Espinosa-Pérez, J. Espinoza, J. Ferguson, P. Gandhi, R. Gao, E. García-Canseco, E. Godoy, M. Hernández-Gómez, M. Hilairet, D. Hissel, I. Husain, A. Jaafar, B. Jayawardhana, D. Jeltsema, S. Junco, F. Kazi, F. Lamnabhi-Lagarrigue, Z. Liu, F. Mancilla-David, R. Mehra, E. Mendes, R. H. Middleton, N. Monshizadeh, P. Monshizadeh, M. Pérez, M. Pirro, A. Pyrkin, S. Sánchez, S. Satpute, J. Scherpen, B. Siciliano, M. Singh, H. Su, R. Talj, E. Tedeschi, A. van der Schaft, D. Wu, M. Zhang, D. Zonetti, for several stimulating discussions and for their hospitality while visiting their institutions.

Some of the topics of this book have been taught by the first author at the EECI Graduate School on Control in Istanbul, Turkey, in 2016, in the Winter Course of the Mexican Association of Automatic Control in 2016, in the Summer School of the Institute of Control Problems of the Academy of Sciences in Moscow, Russia, in 2017 and in the University of Chile, Santiago, Chile, in 2018. A workshop on this topic was organized in Zhejiang University, Hangzhou, China, in 2017.

A large part of this work would not have been possible without the financial support of several institutions. The first author would like to thank ITMO University in Saint Petersburg, Russia, for having sponsored part of this work and the Instituto Tecnológico Autónomo de México (ITAM) for opening its doors for the continuation of his scientific career in Mexico. The second author wishes to thank the Ecole Doctorale-Sciences et Technologies de l'Information des Télécommunications et des Systèmes (ED-STITS) for having funded his doctoral studies and the ITAM for supporting his research activities. The third author wants to thank the National Council of Science and Technology (CONACyT), the Mexican Secretary of Public Education (SEP), and the University of Groningen for all the support received during his academic career. The fourth author wants to thank the University of Newcastle for supporting his research and academic activities.

Mexico/Groningen/Newcastle

Romeo Ortega
José Guadalupe Romero
Pablo Borja
Alejandro Donaire

Acronyms

AC	alternate current
AMM	assumed modes method
CbI	control by interconnection
CL	controlled Lagrangians
DAC	digital-to-analog converter
DC	direct current
DOF	degree(s)-of-freedom
EL	Euler–Lagrange
FOC	field-oriented control
GAS	globally asymptotically stable
GES	global exponential stability
HVDC	high-voltage direct current
IA	integral action
IDA	interconnection and damping assignment
IISS	integral input-to-state stability
ISS	input-to-state stability
LMI	linear matrix inequality
LTI	linear time-invariant
MDICS	matched disturbance integral controlled system
PBC	passivity-based control
PDE	partial differential equation
PEM	proton exchange membrane
pH	port-Hamiltonian
PD	proportional-derivative
PI	proportional-integral
PID	proportional-integral-derivative
PMSG	permanent magnet synchronous generator

PMSM	permanent magnet synchronous motor
PWM	pulse-width modulation
SPR	strictly positive real
VSR	voltage source rectifiers
VTOL	vertical take-off and landing

Notation

Given a vector $x := [x_1, \dots, x_n]^\top \in \mathbb{R}^n$, the symbol $|x|$ denotes its Euclidean norm, i.e. $|x| = \sqrt{x^\top x}$. We denote the ith element of x as x_i. The ith element of the canonical basis of $\mathbb{R}^n$ is represented by e_i. To ease the readability, column vectors are also expressed as $\text{col}(x_1, \dots, x_n)$.

Consider the matrix $B \in \mathbb{R}^{n \times m}$, then B_i denotes the ith column of B, B^k the kth row of B, and B_{ik} the ikth element of B. Moreover, $B^\top$ denotes the transpose of B. Given a square matrix $A \in \mathbb{R}^{n \times n}$, $\text{sym}(A) := \frac{1}{2}\left(A + A^\top\right)$, $\text{skew}(A) := \frac{1}{2}\left(A - A^\top\right)$. To simplify the notation, we express diagonal matrices as $\text{diag}(a_1, \dots, a_n)$, where a_i are the diagonal elements of the matrix.

The symbol I_n denotes the $n \times n$ identity matrix. The symbol $\lambda_i(A)$ refers to the ith eigenvalue of A. In particular, $\lambda_{\text{max}}(A)$, $\lambda_{\text{min}}(A)$ denote the largest and the smallest eigenvalue of A, respectively. A matrix is said to be positive semidefinite if $A = A^\top$ and $x^\top A x \geq 0$ for all $x \in \mathbb{R}^n$, and is said to be positive definite if the inequality is strict, i.e. $x^\top A x > 0$ for all $x \in \mathbb{R}^n \backslash \{0\}$. A is negative (semi)definite if $-A$ is positive (semi)definite. For a positive definite matrix $A \in \mathbb{R}^{n \times n}$ and a vector $x \in \mathbb{R}^n$, we denote the weighted Euclidean norm as $\|x\|_A := \sqrt{x^\top A x}$. The notation used for constant matrices is directly extended to the nonconstant case.

Unless something different is stated, all the functions treated in this book are assumed to be smooth. Moreover, the symbol t is reserved to express time, where we assume $t \in \mathbb{R}_+$. Then, given a function $y : \mathbb{R}_+ \to \mathbb{R}^n$ that depends on time, the symbol $\dot{y}$ denotes the differentiation with respect to time of $y(t)$, i.e. $\dot{y}(t) := py(t)$ where $p = \frac{d}{dt}$. The $\mathcal{L}_\infty$ and $\mathcal{L}_2$ norms of signals are denoted $\|\cdot\|_\infty$ and $\|\cdot\|_2$, respectively.

Given a function $f : \mathbb{R}^n \to \mathbb{R}$ and a vector $x \in \mathbb{R}^n$, we define the differential operator $\nabla_x f(x) := \left(\frac{\partial f}{\partial x}\right)^\top$ and $\nabla_x^2 f(x) := \frac{\partial^2 f}{\partial x^2}$. For a function $F : \mathbb{R}^n \to \mathbb{R}^m$, we define the ijth element of its $n \times m$ Jacobian matrix as $(\nabla_x F(x))_{ij} := \frac{\partial F_j(x)}{\partial x_i}$. When it is clear from the context, we omit the subindex of ∇. Given a distinguished element $x^\star \in \mathbb{R}^n$, we define the matrix $F^\star := F(x^\star) \in \mathbb{R}^{n\times m}$.

Throughout this book, we consider nonlinear systems described by differential equations of the form

$$\Sigma : \begin{cases} \dot{x} &= f(x) + g(x)u, \\ y &= h(x) + j(x)u, \end{cases} \tag{1}$$

where $x(t) \in \mathbb{R}^n$ is the state vector, $u(t) \in \mathbb{R}^m$, $m \leq n$, is the control vector, $y(t) \in \mathbb{R}^m$ is an output of the system defined via the mappings $h : \mathbb{R}^n \to \mathbb{R}^m$ and $j : \mathbb{R}^n \to \mathbb{R}^{m\times m}$, $f : \mathbb{R}^n \to \mathbb{R}^n$ and $g : \mathbb{R}^n \to \mathbb{R}^{n\times m}$ is the input matrix, which is full rank. In the sequel, we will refer to this system as Σ or (f, g, h, j) system.

We also consider the case of port-Hamiltonian systems when the vector field $f(x)$ may be factorized as

$$f(x) = [\mathcal{J}(x) - \mathcal{R}(x)]\, \nabla H(x), \tag{2}$$

where $H : \mathbb{R}^n \to \mathbb{R}$ is the Hamiltonian, $\mathcal{J} : \mathbb{R}^n \to \mathbb{R}^{n\times n}$ and $\mathcal{R} : \mathbb{R}^n \to \mathbb{R}^{n\times n}$, with $\mathcal{J}(x) = -\mathcal{J}^\top(x)$ and $\mathcal{R}(x) = \mathcal{R}^\top(x) \geq 0$, are the interconnection and damping matrices, respectively. To simplify the notation in the sequel, we define the matrix $F : \mathbb{R}^n \to \mathbb{R}^{n\times n}$,

$$F(x) := \mathcal{J}(x) - \mathcal{R}(x).$$

1

Introduction

Motivated by current practice, in this book, we explore the possibility of applying the industry-standard proportional-integral-derivative (PID) controllers to regulate the behavior of nonlinear systems. As is well known, PID controllers are universal, in the sense that they incorporate knowledge of the system's past, present, and future, and they are overwhelmingly dominant in engineering practice. PIDs are highly successful when the main control objective is to drive a given output signal to a constant value. PIDs, however, have two main drawbacks, first, the task of tuning the gains is far from obvious when the system's operating region is large; second, in some practical applications, the control objective cannot be captured by the behavior of output signals.

In this book we show that, for a wide class of systems, these two difficulties can be overcome by exploiting the property of passivity, which in the case of physical systems captures the universal feature of energy conservation. To achieve this end, we propose a new class of controllers called PID passivity-based controls (PBCs), whose main construction principle is to wrap the PID around a passive output of the plant. Since PIDs define (output strictly) passive systems for all positive gains, and the feedback interconnection of passive systems is stable, the proposed architecture yields a highly robust design that preserves stability for all tuning gains – considerably simplifying the task of commissioning the controller. To enable potential designers to use PID-PBCs, we present in the book a comprehensive coverage of this topic.

Since passivity for physical systems is simply a reformulation of energy balancing, it is possible in many practical examples to easily identify some passive outputs. However, in many examples, either these outputs are not the ones we would like to regulate, and/or their desired value is not equal to zero. To address the first problem, we explore in the book the possibility

PID Passivity-Based Control of Nonlinear Systems with Applications, First Edition.
Romeo Ortega, José Guadalupe Romero, Pablo Borja, and Alejandro Donaire.

Published 2021 by John Wiley & Sons, Inc.

of adding an integral action to nonpassive outputs preserving some stability properties. For the second problem, we propose to wrap the PID around the error of the output signal, and then we investigate whether the system is passive with respect to this error signal – a property called *shifted passivity*.

Another scenario of practical interest is when the control objective is to drive the full system *state* to a desired constant value. A classical example is mechanical systems, whose passive outputs are the actuated velocities, but in many applications – e.g. robotics – the objective is to drive *all positions* to some desired constant values. To formulate mathematically this objective, we aim at achieving *Lyapunov stability* of the desired equilibrium, a task that entails the need to construct a Lyapunov function, i.e., a nonincreasing function of the state with a minimum at the desired equilibrium. The approach we adopt in the book to solve this new task is to identify passive outputs whose *integral* can be expressed as a function of the system's state. The identification of these outputs boils down to finding first integrals for the closed-loop dynamics that, in its turn, requires the solution of partial differential equations. The design is completed by projecting the closed-loop dynamics onto the invariant manifold defined by the first integrals and verifying that the resulting function, which depends only on the state of the system, is positive definite.

The aforementioned integrability conditions can be obviated if the system is shifted passive, with a storage function that is positive definite with respect to a desired equilibrium. In spite of the intensive efforts to characterize the class of passive systems that are also shifted passive, the currently available results are quite restrictive – two common requirements being, for instance, that the input matrix is constant and the storage function of the original system is convex. The main results along these lines of research are reviewed in the book.

Another control scenario that we consider in the book pertains to the case when some stabilizing controller has already been added to the system, but we would like to include an additional *integral action* to reject the effect of additive disturbances that were neglected in the design of the aforementioned stabilizing controller. We treat the cases of constant or time-varying disturbances as well as the scenario where the disturbances enter into the image of the input matrix – called matched disturbances – or when they are unmatched. In all cases, we give constructive solutions to the problem of designing this new integral action (IA).

Motivated by the application of PID in physical systems, we pay particular attention in the book to port-Hamiltonian (pH) systems. It is well known that pH models describe the behavior of many physical processes and have the central feature of underscoring the importance of the energy function, the interconnection pattern, and the dissipation of the system, which are the essential ingredients of PBC. As shown throughout the book, the possibility of exploiting the latter features of pH systems in the design of a PID-PBC or an IA allows us to obtain sharper, and in many cases, more constructive results than the ones available for general nonlinear systems.

The book is organized as follows. In Chapter 2, we give the general framework of PID-PBC and show how they can be used to solve several stabilization and output regulation problems. In this chapter, we also discuss some obstacles for the application of PID-PBC, in particular the dissipation obstacle, and show the relationship of PID-PBC with the well-known, and conceptually appealing, control-by-interconnection technique.

In Chapter 3, we show, via several practical examples, how the concept of passivity can be used for the analysis and tuning of PIDs.

In Chapter 4, we discuss the problem of designing of PID-PBCs for the case when the desired value for the regulated output is different from zero. Some results on shifted passivity are presented and their use to solve this problem is discussed. This chapter is wrapped-up with four modern practical applications.

Chapter 5 is devoted to the characterization of passive outputs for pH systems. This result is then used in Chapter 6 to design PID-PBCs for pH systems that ensure Lyapunov stability. The particular case of underactuated mechanical systems is discussed in Chapter 7, where the results are illustrated with several practical examples.

In the final Chapter 8, we present the results pertaining to robustification of nonlinear controllers via the addition of IA – again, paying particular attention to pH and mechanical systems.

The appendices provide the basic definitions and background theory that is used throughout the book. In particular, some preliminaries on passivity and stability theory of state-space systems and a brief discussion on pH systems are given. Some additional technical results and some useful lemmas needed in the book are also presented.

2

Motivation and Basic Construction of PID Passivity-Based Control

In this chapter, we present the PID controller structure considered throughout the book and discuss the motivation to wrap the controller around a passive output – that we call in the sequel PID-PBC. We present the basic construction of PID-PBCs using the so-called *natural passive output*. We discuss the technical issue of well posedness of the feedback interconnection and discuss the role of the dissipation structure of the system on the feasibility of using this kind of PID-PBC. Finally, we briefly discuss the connection of PID-PBC with the formally appealing method of control by interconnection (CbI).

2.1 $\mathcal{L}_2$-Stability and Output Regulation to Zero

Throughout the book, we consider the nonlinear system Σ described in (1) wrapped around a PID controller Σ_c described by

$$\Sigma_c \begin{cases} \dot{x}_c &= y \\ u &= -K_P y - K_I x_c - K_D \dot{y}, \end{cases} \tag{2.1}$$

where $K_P, K_I, K_D \in \mathbb{R}^{m\times m}$ with $K_P, K_I > 0$, and $K_D \geq 0$ are the PID tuning gains. The key property of PID controllers that we exploit in the book is that it defines an *output strictly passive* map $\Sigma_c : y \mapsto (-u)$. This well-known property (Ortega and García-Canseco, 2004; van der Schaft, 2016) is summarized in the lemma below.

Lemma 2.1: *Consider the PID controller* (2.1). *The operator* $\Sigma_c : y \mapsto (-u)$ *is output strictly passive. More precisely, there exists* $\beta \in \mathbb{R}$ *such that the following inequality holds:*

$$\int_0^t y(s)(-u(s))ds \geq \lambda_{\min}(K_P)\int_0^t |y(s)|^2 ds + \beta,\ \forall t \geq 0,$$

PID Passivity-Based Control of Nonlinear Systems with Applications, First Edition.
Romeo Ortega, José Guadalupe Romero, Pablo Borja, and Alejandro Donaire.

Published 2021 by John Wiley & Sons, Inc.

Proof. To prove the lemma, we compute

$$
\begin{aligned}
y^\top(-u) &= \|y\|_{K_P}^2 + y^\top K_I x_c + y^\top K_D \dot{y} \\
&\geq \lambda_{\min}(K_P)|y|^2 + \dot{x}_c^\top K_I x_c + y^\top K_D \dot{y}.
\end{aligned}
$$

Integrating the expression above, we get

$$
\begin{aligned}
&\int_0^t y(s)(-u(s))ds \\
&\qquad \geq \lambda_{\min}(K_P) \int_0^t |y(s)|^2 ds - \|x_c(0)\|_{K_I}^2 - \|y(0)\|_{K_D}^2, \ \forall t \geq 0.
\end{aligned}
$$

The proof is completed setting $\beta := -\|x_c(0)\|_{K_I}^2 - \|y(0)\|_{K_D}^2$. □

The main idea of PID-PBC is to exploit the passivity property of PIDs and, invoking the Passivity Theorem, see Section A.2 of Appendix A, wrap the PID around a *passive output* of the system Σ to ensure $\mathcal{L}_2$-stability of the closed-loop system. This result is summarized in the proposition below, whose proof follows directly from the Passivity Theorem, passivity of the mapping $\Sigma : u \mapsto y$ and output strict passivity of the mapping defined by the PID-PBC.

Proposition 2.1: *Consider the feedback system depicted in Figure 2.1, where Σ is the nonlinear system (1), Σ_c is the PID controller of* (2.1) *and $d(t)$ is an external signal. Assume the interconnection is well defined.*[1] *If the mapping $\Sigma : u \mapsto y$ is passive, the operator $d \mapsto y$ is $\mathcal{L}_2$-stable. More precisely, there exists $\beta \in \mathbb{R}$ such that*

$$
\int_0^t |y(s)|^2 ds \leq \frac{1}{\lambda_{\min}(K_P)} \int_0^t |d(s)|^2 ds + \beta, \ \forall t \geq 0.
$$

Remark 2.1: From Proposition 2.1, we have that, for all signals $d(t) \in \mathcal{L}_2$, the output $y(t) \in \mathcal{L}_2$. Under some additional assumptions, it also follows that $\lim_{t\to\infty} y(t) = 0$. For instance, the latter property holds true, with $d(t) = 0$, under the very weak assumption that there exists a steady-state with u and y constants. Invoking Theorem A.2 it follows that convergence of the output to zero is also achieved if Σ is passive with a positive definite (with respect to

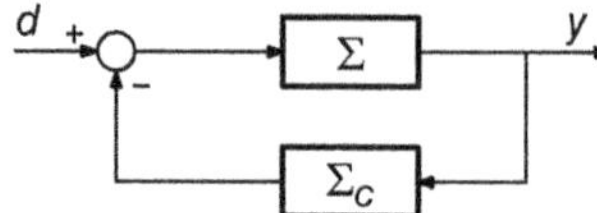

Figure 2.1 Block diagram representation of the closed-loop system of Proposition 2.1.

1 Essentially, it is necessary to ensure that the control law (2.1) can be computed without differentiation nor singularities. We will elaborate on this issue in Section 2.2.

the origin) storage function and the output y is zero-state detectable. Also, we note that the use of passivity in the design can potentially produce, in some cases, slow responses that cannot be improved due to limited tuning parameters, see, for instance, Section 4.5.

Remark 2.2: There are many important practical issues regarding PIDs that are not discussed in the book. For instance, the need to introduce an anti-windup mechanism in the presence of saturation, limitations of the derivative gain, filtering the derivative action, and the use of other architectures of the PID. Also, the simplicity of the PID structure imposes limitations, for example on the set of unstable plants that can be stabilized with this class of controller and the difficulty of handling time delays and complex systems. For a detailed discussion of these, and other PID implementation issues, the interested reader is referred to Ang et al. (2005), Åstrom and Hägglund (1995, 2001), and Åstrom (2018).

2.2 Well-Posedness Conditions

As discussed earlier, it is necessary to ensure that the control law (2.1) can be computed without differentiation nor singularities. The latter may arise due to the presence of the derivative term $\dot{y}$. Clearly, this term can be added only when the output y has relative degree one, that is, when $j(x) = 0$.[2] But even for systems without a derivative action, a singularity may appear for systems with relative degree zero, that is with $j(x) \neq 0$.

The required well-posedness assumptions in both cases are stated in the lemma below, whose proof follows immediately computing the expressions of u for the closed-loop system. As will become clear below, the assumptions are technical only, and rather weak, but they are given for the sake of completeness.

Lemma 2.2: *If the system Σ has relative degree zero, that is $j(x) \neq 0$, the feedback system of Figure 2.1 with $K_D = 0$ is well-posed if the matrix*

$$\mathcal{K}_0(x) := I_m - K_P j(x)$$

is full-rank. On the other hand, if the system Σ has relative degree one, that is $j(x) = 0$ and $K_D > 0$, the feedback system is well posed if the matrix

$$\mathcal{K}_1(x) := I_m - K_D [\nabla h(x)]^T g(x)$$

is full-rank.

2 We recall that a necessary condition for passivity of the system Σ is that the relative degree is smaller or equal to one (van der Schaft, 2016).

Remark 2.3: As a final comment of this section, we note that in van der Schaft (2016), PID control is viewed from a different perspective. Namely, assuming that $\dot{y}$ is *computable*, it is shown that the closed-loop system can be represented as a pH system with *algebraic constraints*. However, leaving aside the complexity of computing $\dot{y}$, the stability analysis of this kind of systems remains an essentially open question.

2.3 PID-PBC and the Dissipation Obstacle

In this section, we reveal a subtle aspect of the practical application of PID-PBC, namely that for passive systems of relative degree one, there exists a steady state *only if* the energy extracted from the controller is zero at the equilibrium. The latter condition is known in PBC as *dissipation obstacle* and is present in many physical systems, for instance, all electrical circuits with leaky energy storing elements operating in nonzero equilibria – i.e. capacitors in parallel, or inductors in series, with resistors. Interestingly, this obstacle is absent in position regulation of mechanical systems since dissipation (due to Coulomb friction) is zero at standstill.

After briefly recalling the nature and mathematical definition of the dissipation obstacle, we prove the claim of inexistence of equilibria stated above in a more general context than just PID-PBC, namely for all dynamic controllers incorporating an integral action on a passive output of relative degree one.

2.3.1 Passive Systems and the Dissipation Obstacle

To mathematically define the dissipation obstacle of a passive system with storage function $S(x)$, let us compute its derivative

$$\dot{S} = [\nabla S(x)]^\top f(x) + [\nabla S(x)]^\top g(x)u. \tag{2.2}$$

As a corollary of Hill–Moylan's theorem, see Theorem A.1, we see that the only passive output of relative degree one is the so-called *natural output*, that we identify with the subindex $(\cdot)_0$, and is given by

$$y_0 = g^\top(x)\nabla S(x). \tag{2.3}$$

Substituting the definition above in (2.2), we can give to it the interpretation of power-balance equation, where $S(x)$ is the energy stored by the system, $y_0^\top u$ is the supplied power and $[\nabla S(x)]^\top f(x)$ is the system's dissipation. In passivity theory, it is said that the system Σ *does not* suffer from the dissipation obstacle – at an assignable equilibrium $x^\star \in \mathcal{E}$ – if

$$[(\nabla S)^\star]^\top f^\star = 0. \tag{2.4}$$

Notice that for pH systems, see Definition D.1, the dissipation obstacle translates into

$$\|(\nabla H)^\star\|^2_{\mathcal{R}^\star} \neq 0, \tag{2.5}$$

where $\mathcal{R}(x)$ is the dissipation matrix and $H(x)$ is a *bona fide* energy function – yielding a clear physical interpretation.

The dissipation obstacle is a phenomenon whose origin is the existence of pervasive dissipation, that is, dissipation that is present even at the equilibrium state. It is a multifaceted phenomenon that has been discussed at length in the PBC literature, where it is shown that the key energy shaping step of PBC (Ortega et al., 2008, Proposition 1), the generation of Casimir functions for CbI (van der Schaft, 2016, Remark 7.1.9) and the assignment of a minimum at the desired point to the shaped energy function (Zhang et al., 2015, Proposition 2) are all stymied by the dissipation obstacle.

2.3.2 Steady-State Operation and the Dissipation Obstacle

The proposition below shows that the application of PID-PBC with the natural output is severely stymied by the dissipation obstacle. Actually, we will prove a much more general result that contains, as a particular case, the PID-PBC scenario.

Proposition 2.2: *Consider the system* Σ *with passive output* (2.3) *and a dynamic extension*

$$\begin{aligned}\dot{x}_c &= g^\top(x)\nabla S(x)\\ \dot{\xi} &= f_c(\xi, x, x_c, u),\end{aligned}$$

where $\xi \in \mathbb{R}^{n_\xi}$. *Define the overall system dynamics as* $\dot{\chi} = F(\chi, u)$, *where* $\chi := \mathrm{col}(x, x_c, \xi)$. *A* necessary *condition for the existence of a constant solution to the equilibrium equation*

$$F(\chi^\star, u^\star) = 0,$$

is that the system Σ *does not suffer from the dissipation obstacle, i.e.,* (2.4) *holds.*

Proof. The proof is established as follows:

$$F(\chi^\star, u^\star) = 0 \Rightarrow \begin{bmatrix}\dot{x}\\ \dot{x}_c\end{bmatrix} = 0.$$

Now

$$\dot{x} = 0 \Rightarrow \dot{S} = 0 \Leftrightarrow [(\nabla S)^\star]^\top f^\star + [(\nabla S)^\star]^\top g^\star u^\star = 0.$$

On the other hand,

$$\dot{x}_c = 0 \Leftrightarrow [(\nabla S)^\star]^\top g^\star = 0.$$

The proof is completed substituting the last identity in the one above. □

Remark 2.4: An immediate corollary of Proposition 2.2 is that the dissipation obstacle hampers the application of PID-PBC for nonzero equilibrium with the natural passive output.

Remark 2.5: As shown in Ortega et al. (2008), van der Schaft (2016), Venkatraman and van der Schaft (2010), and Zhang et al. (2015), one way to overcome the dissipation obstacle is to generate *relative degree zero* outputs. However, it is then not possible to add a derivative term to the controller that, due to its "prediction-like" feature, is useful in some applications. In Chapter 6, we propose a new construction of PID-PBC, where it is possible to add a derivative action to systems where the dissipation obstacle is present. We will also show that the integral and derivative terms perform the energy-shaping process, while the proportional term completes the PBC design by injecting damping into the closed-loop system.

2.4 PI-PBC with y_0 and Control by Interconnection

In this section, we give an interpretation of proportional-integral (PI) PBC with the natural output y_0 as a particular case of CbI, which is a physically (and conceptually) appealing method to *stabilize equilibria* of nonlinear systems widely studied in the literature, cf, Duindam et al. (2009), Ortega et al. (2008), and van der Schaft (2016).

CbI has been mainly studied for pH systems, where the physical properties can be fully exploited to give a nice interpretation to the control action, viewed not with the standard signal-processing viewpoint, but as an energy exchange process. Here, we present CbI in the more general case of the (f, g, h, j)-system Σ, which we assume passive with storage function $S(x)$, and the controller

$$\begin{aligned}
\dot{x}_c &= f_c(x_c) + g_c(x_c)u_c \\
y_c &= h_c(x_c),
\end{aligned}$$

with $x_c(t), u_c(t), y_c(t) \in \mathbb{R}^m$, which is also passive with storage function $S_c(x_c)$. Clearly, the integral action of the PI-PBC is a particular case of this controller with the choices $f_c(x_c) = 0$, $g_c(x_c) = I_m$, and $h_c(x_c) = K_I x_c$.

These systems are coupled via an interconnection that preserves power, that is which satisfies $u^\top y + u_c^\top y_c = 0$. For instance, the classical negative feedback interconnection

$$\begin{aligned} u &= -y_c \\ u_c &= y_0. \end{aligned}$$

The proportional action of the PI-PBC may be assimilated as a preliminary *damping injection* to the plant giving rise to the new process model

$$\dot{x} = [f(x) - g(x)K_P g^T(x)\nabla S(x)] + g(x)u_c.$$

In view of the passivity properties, the storage function of the overall system

$$S_{cl}(x, x_c) := S(x) + S_c(x_c), \tag{2.6}$$

is nonincreasing, alas, not necessarily positive definite – with respect to the desired equilibrium $(x^\star, x_c^\star)$. To construct a *bona-fide* Lyapunov function, it is proposed in CbI to prove the existence of an *invariant* foliation

$$\mathcal{M}_\kappa := \{(x, x_c) \in \mathbb{R}^n \times \mathbb{R}^m \mid x_c = \gamma(x) + \kappa\},$$

with $\gamma : \mathbb{R}^n \to \mathbb{R}^m$ a smooth mapping and $\kappa \in \mathbb{R}^m$. In CbI, a cross-term of the form $\Phi(x_c - \gamma(x) - \kappa)$, with $\Phi : \mathbb{R}^m \to \mathbb{R}$ a *free* differentiable function, is added to the function $S_{cl}(x, x_c)$ given in (2.6) to create the function

$$S_d(x, x_c) := S_{cl}(x, x_c) + \Phi(x_c - \gamma(x) - \kappa),$$

that, due to the invariance property of $\mathcal{M}_\kappa$, satisfies $\dot{S}_d = \dot{S}_{cl}$, hence, is still nonincreasing. If we manage to prove that $S_d(x, x_c)$ is positive definite, the desired equilibrium will be stable. However, the asymptotic stability requirement, and the fact that $\mathcal{M}_\kappa$ is invariant, imposes the constraint on the initial conditions

$$\mathcal{D} = \{x \in \mathbb{R}^n \mid \gamma(x) - x_c = \gamma(x^\star) - x_c^\star\}.$$

That is, the trajectory should start on the leaf of $\mathcal{M}_\kappa$ that contains the desired equilibrium – fixing the initial conditions of the controller. Invoking Sard's theorem (Spivak, 1995), we see that $\mathcal{D}$ is a nowhere dense set, hence, the asymptotic stability claim is *nonrobust* (Ortega, 2021). Two solutions to alleviate this problem – estimation of the constant $\kappa^\star := x_c^\star - \gamma(x_p^\star)$ or breaking the invariance of $\mathcal{M}_\kappa$ via damping injection – have been reported in Castaños et al. (2009), but this adds significant complications to the scheme.

In Chapter 6, we give a solution to the robustness problem using the PI-PBC. In this case, instead of adding the cross term, we project the function $S_{cl}(x, x_c)$ of (2.6) onto $\mathcal{M}_\kappa$ to generate the function:

$$V(x) := S_{cl}(x, \gamma(x) + \kappa),$$

that we might be able to use as a Lyapunov function (for the projected dynamics). To avoid the difficulty of the constraint on the initial conditions mentioned above, we replace the PI by a *static* state-feedback law of the form

$$u = -K_P y_0 - K_I[\gamma(x) - \gamma(x^\star)],$$

which exactly coincides with the control generated with the PI-PBC evaluated at $\mathcal{M}_\kappa^\star$. Such an implementation does not impose any constraint on $x_c(0)$ and is therefore robust.

The details of this construction are given in Chapter 6, where we also prove that the set of solutions of the partial differential equations (PDEs) that must be solved to generate the invariant foliation $\mathcal{M}_\kappa$ in CbI is *strictly* smaller than the ones needed in the PID-PBC, yielding a design procedure that is applicable for a broader class of systems.

Bibliography

K. H. Ang, G. Chong, and Y. Li. PID control system analysis, design and technology. *IEEE Transactions on Control Systems Technology*, 13(4): 559–576, 2005.

K. J. Åstrom. Advances in PID control. In *XXXIX Jornadas de Automatica*, Badajoz, Spain, 2018.

K. J. Åstrom and T. Hägglund. *PID Controllers: Theory, Design, and Tuning*. 2nd edition. Instrument Society of America, 1995.

K. J. Åstrom and T. Hägglund. The future of PID control. *Control Engineering Practice*, 9(11): 1163–1175, 2001.

F. Castaños, R. Ortega, A. J. van der Schaft, and A. Astolfi. Asymptotic stabilization via control by interconnection of port-Hamiltonian systems. *Automatica*, 45(7): 1611–1618, 2009.

V. Duindam, A. Macchelli, S. Stramigioli, and H. Bruyninckx. *Modeling and Control of Complex Physical Systems: The Port-Hamiltonian Approach*. Springer Science & Business Media, 2009.

R. Ortega and E. García-Canseco. Interconnection and damping assignment passivity-based control: a survey. *European Journal of Control*, 10(5): 432–450, 2004.

R. Ortega, A. J. van der Schaft, F. Castaños, and A. Astolfi. Control by interconnection and standard passivity–based control of port–Hamiltonian systems. *IEEE Transactions on Automatic Control*, 53(11): 2527–2542, 2008.

R. Ortega. Comments on recent claims about trajectories of control systems valid for particular initial conditions. *Asian Journal of Control*, 1–8, 2021.

M. Spivak. *Calculus on Manifolds*. Addison-Wesley, 1995.

A. J. van der Schaft. *L_2-Gain and Passivity Techniques in Nonlinear Control*. Springer-Verlag, Berlin, 3rd edition, 2016.

A. Venkatraman and A. J. van der Schaft. Energy shaping of port-Hamiltonian systems by using alternate passive input-output pairs. *European Journal of Control*, 16(6): 665–677, 2010.

M. Zhang, R. Ortega, D. Jeltsema, and H. Su. Further deleterious effects of the dissipation obstacle in control-by-interconnection of port-Hamiltonian systems. *Automatica*, 61(11): 227–2331, 2015.

3

Use of Passivity for Analysis and Tuning of PIDs: Two Practical Examples

As discussed in Chapter 1, PIDs guarantee a good steady-state behavior in most engineering applications. However, the stringent quality requirements imposed in current times demand to ensure also a good *transient performance*. This requirement becomes particularly important in applications where the range of operation of the system is wide for the following two reasons. On the one hand, the wider the operating range, the more conspicuous become the nonlinear effects of the system's dynamics, invalidating the use of linear models to describe their behavior. On the other hand, the task of tuning the PID gains, which almost always relies on the use of linear models and linear analysis tools – like pole-zero locations or frequency response considerations – becomes particularly difficult. Some popular techniques to overcome this problem, including their well-known shortcomings, are discussed in Chapter 1 and references therein.

In this chapter, we illustrate with two practical applications – control of induction motors and regulation of the air supply of a fuel cell system – how the property of passivity can be used to select the gains of the PIDs without appealing to linearization-based techniques. A key step in both applications is the decomposition of the system as a feedback interconnection of two subsystems amenable for their passivity characterization. In the first application, the gain tuning is complicated by the fact that a critical parameter of the system, i.e. the rotor resistance, is highly uncertain and changes with temperature. Therefore, we concentrate our attention on defining intervals for the PID gains where stability is preserved for the largest possible range of variation of this parameter – clearly, the aforementioned intervals are those for which the required passivity property is preserved.

For the second application, we study the control of the air subsystem that feeds the fuel cell cathode with oxygen – whose dynamics is described with

PID Passivity-Based Control of Nonlinear Systems with Applications, First Edition.
Romeo Ortega, José Guadalupe Romero, Pablo Borja, and Alejandro Donaire.

Published 2021 by John Wiley & Sons, Inc.

a widely accepted nonlinear model. Due to the complexity of this model, the model-based controllers that have been proposed for this application are designed using its linear approximation at a given equilibrium point, which might lead to conservative stability margin estimates for the usually wide operating ranges of the system. On the other hand, practitioners propose the use of simple proportional or PI controllers around the compressor flow, which ensures good performance in most applications. In this chapter, we provide the theoretical justification to this scheme, proving that this output variable has the remarkable property that the linearization (around any admissible equilibrium) of the input-output map is *strictly passive*. Hence, the controllers used in applications yield (locally) asymptotically stable loops – for any desired equilibrium point and *all values of the controller gains*. Ensuring stability for all tuning gains overcomes the inherent conservativeness of linearized dynamics analysis and assures the designer on the current use of robust, high-performance loops. Instrumental to prove the passivity property is the exploitation of some monotonicity characteristics of the system that stem from physical laws.

3.1 Tuning of the PI Gains for Control of Induction Motors

Due to its high reliability field-oriented control (FOC) is the standard for high dynamic performance induction motor drives. Historically, this remarkable controller was derived proceeding from physical intuition and a deep understanding of the machine operation, with little concern about a rigorous analytical study of its stability and performance, see Blaschke (1972). An approximate analysis (based on steady-state behavior, time-scale assumptions, and linearizations, e.g. Bose (1986) and Leonhard (1985)) can be combined with the designer expertise to commission the controller in simple applications. However, to meet large bandwidth requirements, or other tight specifications, this ad hoc commissioning stage may be time-consuming and expensive. To simplify the off-line tuning of FOC, and eventually come to terms with its achievable performance, a better theoretical understanding of the dynamic behavior of FOC is unquestionably needed. Such an analysis is unfortunately stymied by the fact that the dynamic behavior of the closed loop is described by complex nonlinear relationships.

Realizing the practical importance of FOC and motivated by the need to clarify its theoretical underpinnings, a lot of research has been carried out on this topic in the last few years. First, in Ortega et al. (1996), we proved that FOC has a very nice energy-dissipation interpretation, which is expressed in terms of passivity of certain subsystems – this result provides a deep system-theoretic foundation to this popular strategy and paves the way for subsequent analysis. We also proved that the passivity-based controller that we proposed for voltage-fed machines exactly reduces to FOC for current-fed machines. A corollary of this result is the first rigorous proof of global asymptotic stability of FOC reported in Ortega and Taoutaou (1996a). Later on, in de Wit et al. (1996), we established the stronger property of global *exponential* stability (GES), and proved as a by-product some remarkable robust stability properties of FOC with respect to uncertainty in the rotor time-constant, see also Reginatto and Bazanella (2003). In Kim et al. (1997), we carried out a theoretical and experimental comparison of FOC and feedback linearization, which unquestionably proved the superiority of the former. Motivated by practical considerations we developed a *discrete-time* version of FOC with guaranteed stability properties in Ortega and Taoutaou (1996a). It was later tested experimentally in Taoutaou et al. (1997). Other related works, which have been reported within the industrial electronics community may be found in Cecati (1997). See also Nicklasson et al. (1997) where we have extended our work to the general rotating machine. The interested reader is referred to Marino et al. (2010) and Ortega et al. (1998) for a tutorial account on modern control of induction motors, including FOC.

In this section, we continue with this line of research and concentrate on the practically important problem of *off-line tuning* of the gains of the PI speed loop. It is well known that the performance of FOC critically depends on the tuning of these gains, a task which is rendered difficult by the high uncertainty on the rotor time-constant. We give here some simple rules to carry out the PI tuning ensuring *robust* stability with respect to this parameter. The main contribution of our work is to propose a simple frequency response test that, for each setting of PI gains, estimates the maximum error of the rotor time-constant for which *global* stability is guaranteed. In this way, we can estimate the *stability margin* of a PI controller before closing the loop. The interplay between robust stability (i.e. stable operation in large ranges of parameter uncertainty) and performance (in the classical sense of fast-transient responses with small overshoot) is far from being obvious, particularly in a nonlinear setting. It is clear, however, that some margin of stability must be assured for a correct operation of the system. It is reasonable then to expect that robustly stable tunings would yield "better" performance. In the lack of an alternative way to evaluate performance improvement, this is the presumption made in this section.

In Section 3.1.1, using as a preamble the model of a current-fed induction motor and the equations of the indirect FOC, the problem approached in this section is presented. In Section 3.1.2, an alternative representation for the closed-loop is developed. This representation reveals some new energy dissipation features of FOC; hence, its instrumental for our theoretical developments. The main topic of the section, finding conditions to achieve global stability, is approached in Section 3.1.3. Section 3.1.4 contains some concluding remarks.

Further details of the material reported in this chapter may be found in Chang et al. (2000).

3.1.1 Problem Formulation

The dynamic model of the current-fed induction motor expresses the rotor flux and the stator currents in a reference frame rotating at the rotor angular speed[1]

$$\dot{\lambda}_r = -\frac{1}{T_r}\lambda_r + \frac{L_{sr}}{T_r}i_s, \tag{3.1}$$

$$\dot{\omega}_r = \frac{1}{J}(\tau - \tau_L), \tag{3.2}$$

$$\tau = \frac{n_p L_{sr}}{L_r} i_s^T \mathcal{J} \lambda_r, \tag{3.3}$$

where λ_r is the two-dimensional rotor flux vector, i_s is the two-dimensional stator current vector, ω_r is the rotor speed, τ is the generated torque, $T_r = \frac{L_r}{R_r}$ is the rotor time-constant, L_r is the rotor self-inductance, R_r is the rotor resistance, L_{sr} is the mutual inductance, J is the shaft inertia-moment, n_p is the number of pole-pairs, τ_L is the load torque and

$$\mathcal{J} = \begin{bmatrix} 0 & -1 \\ 1 & 0 \end{bmatrix}.$$

In the above Eqs. (3.1)–(3.3), we assume that R_r and τ_L are constant but unknown, while other parameters are constant and known.

In indirect FOC, the stator current is chosen as

$$i_s = \mathbf{e}^{\mathcal{J}\rho_d} \begin{bmatrix} \frac{\beta^\star}{L_{sr}} \\ \frac{L_r}{n_p L_{sr} \beta^\star}\tau_d \end{bmatrix}, \tag{3.4}$$

where the rotation matrix $\mathbf{e}^{\mathcal{J}\rho_d}$ is

$$\mathbf{e}^{-\mathcal{J}\rho_d} = \begin{bmatrix} \cos(\rho_d) & -\sin(\rho_d) \\ \sin(\rho_d) & \cos(\rho_d) \end{bmatrix},$$

1 See, e.g. Marino et al. (2010) and Ortega et al. (1998), for a derivation of this model from the textbook standard model.

$\beta^{\star} > 0$ is the constant reference value of the rotor flux norm[2] and the angle of the desired rotor flux ρ_d is given by

$$\dot{\rho}_d = \frac{\hat{R}_r}{n_p \beta^{\star 2}} \tau_d \tag{3.5}$$

with the estimated rotor resistance $\hat{R}_r > 0$. The desired torque τ_d is, in speed regulation applications, typically defined via a PI speed loop as

$$\tau_d = -\left(K_P + \frac{K_I}{p}\right)(\omega_r - \omega^{\star}), \tag{3.6}$$

where $p = \frac{d}{dt}$, $\omega^{\star}$ is the speed reference and $K_P, K_I > 0$.

It has been shown in Ortega and Taoutaou (1996a) that the system (3.1)–(3.6) is globally asymptotically stable (GAS) if $\hat{R}_r = R_r$. Furthermore, in de Wit et al. (1996), we proved that the stability is actually *exponential* and showed that the system remains stable under large variations of the rotor resistance. The problem we address in this section is how to choose the parameters K_P and K_I to guarantee, not only stability of the closed-loop but also a good *performance* in spite of the uncertainty on the rotor resistance R_r. To formulate mathematically this problem, we take the standpoint advocated in the introduction, namely that "robust stability implies enhanced performance with good PI tunings." Hence, we look for PI tunings which allow larger rotor resistance estimation errors.

Thus, the problem is formulated as follows:

- Given the induction motor parameters L_r, J, and an *uncertainty interval* for the rotor resistance $R_r \in [R_r^{\min}, R_r^{\max}]$ (clearly containing $\hat{R}_r$). Find a range of PI gains (if one exists) for which global asymptotic stability of the closed-loop system is preserved.

Our analysis can also be used in a dual manner of theoretical interest.

- Given the induction motor parameters L_r, J, an *a-priori* estimate of the rotor resistance $\hat{R}_r$, and a controller setting K_P, K_I. Find a range of values of the relative resistance estimation error

$$\delta \triangleq \frac{\hat{R}_r}{R_r}$$

for which global asymptotic stability of the closed-loop system is preserved. More precisely, we want to find an interval $\Delta \triangleq [\delta^{\min}, \delta^{\max}]$ (clearly containing one) such that, if $\delta \in \Delta$, then the system (3.3)–(3.6) is GAS.

---

2 We have considered constant $\beta^{\star}$ only for ease of presentation. The time-varying case follows *verbatim* from our derivations in Nicklasson et al. (1997).

Even though some answers to these questions may be found in de Wit et al. (1996) and Reginatto and Bazanella (2003) the procedure relies on the generation of Lyapunov functions; Hence, it is not very transparent to the user. In this section, we give a very simple *frequency response*[3] procedure that, for each setting of the controller gains, generates the required set of values for δ. The *size* of this interval, which we will call in the sequel the *performance interval*, provides then a robustness measure of the closed-loop system which guides the user in the choice of the PI gains. We should also underscore that the procedure only requires the knowledge of L_r, J, and is independent of the other motor parameters and the reference values $\beta^\star$ and $\omega^\star$.

3.1.2 Change of Coordinates

In order to solve the problems formulated above, an alternative representation for the closed-loop of the induction motor and the indirect FOC is presented in this section. This representation reveals some new energy dissipation features of FOC which are instrumental for our theoretical developments.[4]

dq-Coordinates

We first express the rotor flux in the rotating reference frame and "rotate back" the stator currents (3.4) as[5]

$$\begin{bmatrix} \lambda_{sd} \\ \lambda_{sq} \end{bmatrix} \triangleq \mathbf{e}^{-\mathcal{J}\rho_d}\lambda_r, \quad \begin{bmatrix} i_{sd} \\ i_{sq} \end{bmatrix} \triangleq \begin{bmatrix} \frac{\beta^\star}{L_{sr}} \\ \frac{L_r}{n_p L_{sr}\beta^\star}\tau_d \end{bmatrix}.$$

Using these coordinates, the motor model (3.1) in closed-loop with the FOC (3.4)–(3.6) yields

$$\dot{\lambda}_{rd} = -\frac{R_r}{L_r}\lambda_{rd} + \frac{\hat{R}_r}{n_p(\beta^\star)^2}\tau_d\lambda_{rq} + \frac{R_r\beta^\star}{L_r}, \tag{3.7}$$

$$\dot{\lambda}_{rq} = -\frac{R_r}{L_r}\lambda_{rq} - \frac{\hat{R}_r}{n_p(\beta^\star)^2}\tau_d\lambda_{rd} + \frac{R_r}{n_p\beta^\star}\tau_d, \tag{3.8}$$

3 The concept of frequency response has a clear physical interpretation and is familiar to the industrial electronics community.

4 The importance of coordinate changes was probably first underscored by Copernicus who pointed out that the planetary motions are better understood from the sun's perspective.

5 To avoid cluttering we keep here the same notation λ, i for fluxes and currents in all coordinates.

$$\dot{\tilde{\omega}}_r = \frac{1}{J}\left[\frac{1}{\beta^\star}\lambda_{rd}\tau_d - \frac{n_p\beta^\star}{L_r}\lambda_{rq} - \tau_L\right], \tag{3.9}$$

$$\dot{\tau}_d = -\frac{K_P}{J}\left[\frac{1}{\beta^\star}\lambda_{rd}\tau_d - \frac{n_p\beta^\star}{L_r}\lambda_{rq} - \tau_L\right] - K_I\tilde{\omega}_r, \tag{3.10}$$

where we have defined the speed error $\tilde{\omega}_r \triangleq \omega_r - \omega^\star$.

State-Space Representation

To further simplify the notation, we also define

$$v = \left[v_1,\ v_2,\ v_3,\ v_4\right]^\top = \left[\lambda_{rq},\ \lambda_{rd},\ \tilde{\omega}_r,\ \tau_d\right]^\top$$

to rewrite (3.7)–(3.10) as

$$\dot{v} = \begin{bmatrix} -\frac{R_r}{L_r} & -\frac{\hat{R}_r}{n_p(\beta^\star)^2}v_4 & 0 & \frac{R_r}{n_p\beta^\star} \\ \frac{\hat{R}_r}{n_p(\beta^\star)^2}v_4 & -\frac{R_r}{L_r} & 0 & 0 \\ -\frac{n_p\beta^\star}{JL_r} & \frac{1}{J\beta^\star}v_4 & 0 & 0 \\ \frac{K_Pn_p\beta^\star}{JL_r} & -\frac{K_P}{J\beta^\star}v_4 & -K_I & 0 \end{bmatrix} v + \begin{bmatrix} 0 \\ \frac{R_r\beta^\star}{L_r} \\ -\frac{1}{J}\tau_L \\ \frac{\tau_L}{J} \end{bmatrix}. \tag{3.11}$$

It is shown in de Wit et al. (1996) that the equilibria of this system (i.e. the values of v for which $\dot{v} = 0$), are defined by very complex algebraic relationships, and the system can actually have multiple equilibria. On the other hand, when $\tau_L = 0$, the equilibrium is unique and is given by

$$[\bar{v}_1, \bar{v}_2, \bar{v}_3, \bar{v}_4]^T = [0, \beta^\star, 0, 0]^T.$$

Given this fact, and the stability consideration mentioned in the previous paragraph, we will treat the load torque as a *disturbance* and concentrate on the case $\tau_L = 0$.

We now *shift the equilibrium* of (3.11) to the origin. To this end, we introduce the change of coordinates $z_i = v_i - \bar{v}_i,\ i = 1, \dots, 4$, and write (3.11) in terms of these new variables as

$$\dot{z} = \begin{bmatrix} -\frac{R_r}{L_r} & -\frac{\hat{R}_r}{n_p(\beta^\star)^2}z_4 & 0 & \frac{R_r-\hat{R}_r}{n_p\beta^\star} \\ \frac{\hat{R}_r}{n_p(\beta^\star)^2}z_4 & -\frac{R_r}{L_r} & 0 & 0 \\ -\frac{n_p\beta^\star}{JL_r} & \frac{1}{J\beta^\star}z_4 & 0 & \frac{1}{J} \\ \frac{K_Pn_p\beta^\star}{JL_r} & -\frac{K_P}{J\beta^\star}z_4 & -K_I & -\frac{K_P}{J} \end{bmatrix} z. \tag{3.12}$$

This is a set of nonlinear ordinary differential equations of the form $\dot{z} = f(z)$ with a unique equilibrium at the origin, whose stability properties we study in the next subsections.

A Remark on Direct FOC

There exists another version of FOC called *direct*, which actually precedes the indirect version discussed above, and has a very simple interpretation. To reveal this fact, let us define the polar coordinates for the rotor flux as

$$\beta := |\lambda_r|, \quad \rho := \arctan\left\{\frac{\lambda_{r2}}{\lambda_{r1}}\right\},$$

and introduce a change of coordinates for the currents

$$\begin{bmatrix} i_d \\ i_q \end{bmatrix} := \mathbf{e}^{-\mathcal{J}\rho} i_s.$$

In polar coordinates (β, ρ), the electrical dynamics of the motor takes the form

$$\begin{aligned}
\dot{\beta} &= -\frac{R_r}{L_r}[\beta - L_{sr} i_d],\\
\dot{\rho} &= \frac{R_r}{n_p \beta^2}\tau,\\
\tau &= \frac{n_p L_{sr}}{L_r}\beta i_q.
\end{aligned}$$

The representation above clearly motivates the choice of the controller

$$i_s = \frac{1}{L_{sr}}\mathbf{e}^{\mathcal{J}\rho}\begin{bmatrix} \beta^\star \\ \frac{L_r}{n_p \beta^\star}\tau_d \end{bmatrix},$$

with τ_d given by (3.6) that yields

$$\begin{aligned}
\dot{\beta} &= -\frac{R_r}{L_r}[\beta - \beta^\star],\\
\tau &= \frac{\beta}{\beta^\star}\tau_d,
\end{aligned}$$

which ensures the control objectives, i.e. $\lim_{t\to\infty}\beta(t) = \beta^\star$ and $\lim_{t\to\infty} \omega(t) = \omega^\star$. The main problem with this controller is that it presumes the knowledge of the rotor angle ρ, which is, usually, not available for measurement. This fact, motivated the replacement of this signal by its "desired" values ρ_d in the indirect FOC (3.4). For additional discussion on FOC, the reader is referred to Ortega et al. (1998) and Marino et al. (2010).

3.1.3 Tuning Rules and Performance Intervals

In this section, we propose a frequency-based procedure to test global stability of the closed-loop system (3.12). This procedure allows us to evaluate the performance interval Δ for each given setting of the PI gains. As mentioned above, the size of these intervals gives a quantitative measure of the robustness of the PI tuning. They can also be used to determine ranges of allowable variations for the PI gains for a given rotor resistance uncertainty interval.

To establish this result, we find it convenient to decompose the closed-loop system (3.12) as the feedback interconnection of two subsystems. One of the subsystems contains all nonlinearities, but turns out to be passive. The second subsystem is linear time-invariant (LTI). Our motivation for introducing this decomposition is twofold. First, as shown in Proposition A.3, it is well-known that the negative feedback interconnection of two passive subsystems is still passive. Furthermore, if one of them is strictly passive, then the closed-loop system is $\mathcal{L}_2$-stable. Second, for LTI systems, there is a very simple frequency-domain characterization of passivity in terms of positivity of the real part of its transfer function. Since this transfer function depends on the motor parameters and the PI gains, this positive realness test will provide us with a criterion for global stability.

Passive Subsystems Decomposition

To obtain the desired feedback decomposition, we rewrite (3.12) "pulling out" its nonlinear part as

$$\dot{z} = \begin{bmatrix} -\frac{R_r}{L_r} & 0 & 0 & \frac{1}{n_p\beta^\star}\left(\hat{R}_r - R_r\right) \\ 0 & -\frac{R_r}{L_r} & 0 & 0 \\ -\frac{n_p\beta^\star}{JL_r} & 0 & 0 & \frac{1}{J} \\ \frac{K_P n_p\beta^\star}{JL_r} & 0 & -K_I & -\frac{K_P}{J} \end{bmatrix} z + \begin{bmatrix} -\frac{\hat{R}_r}{n_p(\beta^\star)^2} z_2 z_4 \\ \frac{\hat{R}_r}{n_p(\beta^\star)^2} z_1 z_4 \\ \frac{1}{J\beta^\star} z_2 z_4 \\ -\frac{K_P}{J\beta^\star} z_2 z_4 \end{bmatrix}. \tag{3.13}$$

We further observe that, to describe the dynamics of z_2, we can define an "LTI system" as

$$\dot{z}_2 = -\frac{1}{T_r} z_2 + \frac{\hat{R}_r}{n_p(\beta^\star)^2} z_1 z_4,$$

and, consequently, represent (3.12) as the feedback interconnection of two LTI subsystems with two multipliers – which capture the nonlinearities of the system – as shown in Figure 3.1. The transfer function of the feedback LTI subsystem is

$$G_2(s) = \frac{R_r\hat{R}_r}{n_p^2(\beta^\star)^2} \frac{\frac{\hat{R}_r}{R_r}s^2 + K_P' s + K_I'}{s^3 + \left(K_P' + \frac{R_r}{L_r}\right)s^2 + \left(K_I' + K_P'\frac{\hat{R}_r}{L_r}\right)s + K_I'\frac{\hat{R}_r}{L_r}}, \tag{3.14}$$

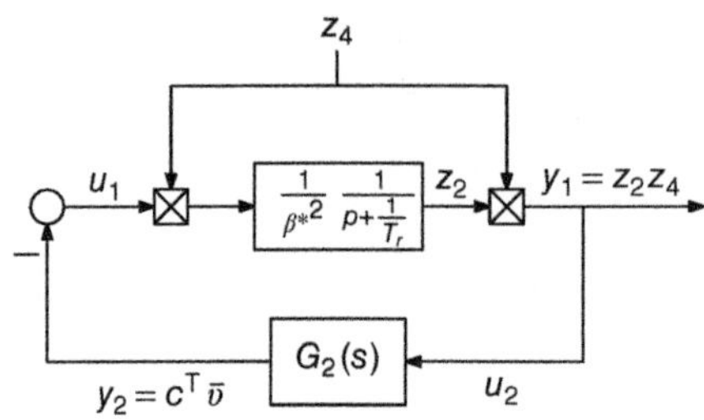

Figure 3.1 Feedback decomposition of the closed-loop system.

where we have defined $K_P' = \frac{K_P}{J}, K_I' = \frac{K_I}{J}$. It is important to underscore that, up to a positive scaling gain, $G_2(s)$ depends only on δ, L_r, $\hat{R}_r$ and the normalized PI gains.

Consequently, the local stability of (3.13) close to the equilibrium point will be determined by the poles of $G_2(s)$. To better illustrate the effect of the parameter uncertainty, it is convenient to rewrite $G_2(s)$ as

$$G_2(s) = \frac{\hat{R}_r^2}{n_p^2(\beta^\star)^2} \frac{s^2 + \frac{K_P'}{\delta}s + \frac{K_I'}{\delta}}{s(s^2 + K_P's + K_i') + \frac{\hat{R}_r}{L_r\delta}(s^2 + K_P'\delta s + K_I'\delta)}. \tag{3.15}$$

Notice that in the ideal case, when $\hat{R}_r = R_r$, $G_2(s)$ reduces to a first-order transfer function with a pole in $-\frac{R_r}{L_r}$. Invoking continuity considerations, we can also conclude that there will be an interval on δ for which local stability is preserved. Local stability is clearly a necessary condition for global stability, henceforth, the first step in the determination of the performance intervals will be the verification of stability of $G_2(s)$, the second step being verifying its positive realness.

Global Stability

The key observation to study the stability of the feedback system of Figure 3.1 is that the (nonlinear) forward subsystem G_1 – with input u_1 and output y_1 – is passive. This fact is proven in Exercise 5 in Chapter VI of (Desoer and Vidyasagar, 2009). Therefore, invoking the passivity theorem and recalling that LTI systems are strictly passive if and only if their transfer function is strictly positive real (SPR), to ensure stability of the closed-loop system it suffices to verify the conditions for $G_2(s)$ to be SPR.

We are in a position to present our main result, whose proof is a direct application of the definition of SPR transfer functions and the passivity theorem. The interested reader is referred to Appendix A in Chang et al. (2000) where a detailed proof of the proposition is given.

Proposition 3.1: *Consider the model of the current-fed induction motor* (3.1)–(3.3) *in closed-loop with the indirect FOC* (3.4)–(3.6). *The closed-loop, with* $\tau_L = 0$, *is* GAS *for all positive values of* δ *such that*

P1 (Local stability)

$$\frac{1}{\delta} > \frac{K_I'}{K_P' \frac{\hat{R}_r}{L_r} + K_I'} - K_P' \frac{L_r}{\hat{R}_r}. \tag{3.16}$$

P2 (SPR)

$$K_P' \frac{L_r}{\hat{R}_r}(\delta - 1) > -1,$$

$$(K_P')^2 + K_I' \left[2\sqrt{1 + K_P' \frac{L_r}{\hat{R}_r}(\delta - 1)} - \left(\delta + \frac{1}{\delta}\right) \right] > 0.$$

When $\tau_L \neq 0$ *all trajectories enter (in finite time) a ball centered at the origin of radius* $|\tau_L|$.

From (3.16), it is easy to see that local stability is preserved – for all PI tuning gains – whenever the rotor resistance is underestimated. Finally, the region in parameter space (δ, K_p', K_I'), where the inequalities above hold corresponds to a global stability region. Standard optimization techniques can be used to precisely define this region.

Performance Intervals

For the estimation of the performance interval, since it is a single parameter search, we can provide the following simple algorithm:

Step 1. Input data: Numerical values for L_r, J, the rotor resistance estimate $\hat{R}_r > 0$ and the controller gains $K_P > 0$, $K_I > 0$. Set $\delta = 1$.

Step 2. Check the inequalities of Proposition 3.1, if one of them is not satisfied then $\delta^{\max}$ is found. Go to **Step 4**. If none of the inequalities is satisfied, then proceed with the following step.

Step 3. Increment of the current values δ by a small positive number and go to **Step 2.**

Step 4. Set $\delta = 1$.

Step 5. Decrease the current value of δ by a small number.

Step 6. Check the inequalities of Proposition 3.1, if one of them is not satisfied then $\delta^{\min}$ is found and the algorithm stops. If both conditions hold, go to **Step 5.**

3.1.4 Concluding Remarks

In this section, we have given some simple rules to tune the PI gains of indirect FOC for induction motors in order to improve the transient performance. The performance enhancement is quantitatively measured with an indicator of robustness of the stability with respect to uncertainty on the rotor resistance. A very simple algorithm that evaluates this indicator (i.e. the performance interval Δ) for each PI setting is presented. Our theoretical results were validated with some simulation and experimental evidence reported in Chang et al. (2000).

It is interesting to remark that a simple condition for local stability is to choose the damping ratio of the speed-loop characteristic polynomial $s^2 + K_P' s + K_I'$ larger than $\frac{1}{2}$ or underestimating the rotor resistance. We have also shown that the speed behavior is improved when the performance interval, in which the global stability condition holds, is increased. The increase of the performance interval is obtained by increasing either the damping ratio or the natural frequency of the same characteristic equation, or both the damping ratio and the natural frequency.

A closing remark should be made concerning the assumption of zero load torque. In the presence of the latter, our analysis ensures only that the error signals enter some ball of radius $|\tau_L|$, and the performance could be degraded. In applications where load torque is significant (i.e. close to the rated load torque), we should appeal to the Lyapunov function construction of (de Wit et al., 1996), where the general case is rigorously treated. If we have some prior knowledge on the load torque range, this will result in an accurate estimate of the performance interval. The price to be paid is that a more complicated optimization algorithm would need to be implemented to find the Lyapunov function coefficients, but these calculations are done offline and only once.

3.2 PI-PBC of a Fuel Cell System

A fuel cell is an electrochemical device in which the energy of a reaction between a fuel (e.g. hydrogen) and an oxidant (e.g. oxygen), is converted directly and continuously into electrical energy. Fuel cell systems offer a clean alternative to energy production and are currently under intensive development. This new device is known for its low operating temperature, few moving parts (mechanically robust construction), and low-to-zero emissions during its operation. In particular, it is considered one of the most promising alternatives to fossil-based fuel engines for automotive applications, where the fuel cell is combined with ultracapacitors or batteries to feed the load. In such hybrid systems, the energy management is assured by the control of DC/DC converters. To predict, monitor and efficiently control the fuel cell system under a variety of environmental conditions and a wide operating range, it is essential to understand the highly nonlinear behavior of the fuel cell itself.

We consider here a proton exchange membrane (PEM) fuel cell system composed of four main subsystems: the hydrogen subsystem that feeds the anode with hydrogen, the air subsystem that feeds the cathode by oxygen, the humidifier, and the cooler that maintain the humidity degree and the temperature of the fuel cell, respectively. Knowing that the degree of humidity and the temperature cannot change rapidly, the control problem of these two subsystems can be decoupled from the rest of the system. On the other hand, the hydrogen subsystem is controlled by an electrical valve, while the air subsystem is controlled with a slower mechanical device (e.g. an electrical motor and a compressor), suggesting another time-scale decomposition. Furthermore, the goal of the hydrogen flow control is to minimize the pressure difference across the membrane, that is, the difference between anode and cathode pressures. Using simple proportional control based on the pressure difference, the pressure in the anode can quickly follow the changes in the cathode pressure. For all these reasons, the control problem is concentrated on the air supply system equipped with an electrical motor and a compressor. The interested reader is referred to (Pukrushpan et al., 2004) for a detailed description of the fuel cell control problem.

The fuel cell presents the problem of oxygen starvation when the load changes rapidly. If the load increases, more power is needed, leading to an increase in the fuel cell current. Then, the chemical reactions accelerate to give the required power to the load, consuming more oxygen. To avoid starvation, the compressor motor should accelerate to feed the fuel cell with the suitable quantity of oxygen. But the motor of the compressor is directly fed by the fuel cell itself, hence, it consumes a part of its power. This fact complicates the control problem of the fuel cell system.

There are two performance variables in compromise. Namely, the net power delivered to the load and the oxygen stoichiometry (ratio of the input oxygen flow over the reacted oxygen flow in the cathode). Previous work presented in Pukrushpan et al. (2004) proves that, for all operating points, the maximum net power delivered to the load is reached approximately for the same constant value of stoichiometry, equal to two. Based on this result, we fixed our control objective: we need to *avoid starvation* by controlling the oxygen stoichiometry around a constant desired value, which corresponds approximately to the *m*aximum net power delivered to the load in steady state. The first constraint is ensured by the design of a suitable dynamic controller, while the second one is satisfied by the choice of the controller reference signal. The control input is the voltage applied to the motor of the compressor, and the disturbance input is the fuel cell current that reflects the load changes.

The dynamic behavior of the air supply system is highly nonlinear and uncertain, which renders the analysis and design of suitable control laws very complicated. In previous works (Talj et al., 2009, 2010), we have studied and compared different control laws to improve the transient performance and avoid unacceptable undershoot of the oxygen excess ratio during transients of current demand. In this study, we aim to get a better understanding of the characteristics of the system. Once the stability limits are known and the system is well understood, it will be possible to extend the previous studies and even to propose new control strategies and control methods.

In Pukrushpan et al. (2004) a ninth order nonlinear model – derived from physical principles – to describe the dynamics of the fuel cell was proposed. Reduced order models of the air supply system were proposed in Talj et al. (2009, 2010). In spite of its widespread acceptance by the scientific community, the complexity of these models stymies its application for controller design, which is almost invariably carried out using its linear approximation at a *given* equilibrium point. The validity of this analysis is restricted to a neighborhood of that particular point, even though the system has a wide operating region. Consequently, the resulting numerical designs may be conservative and may fail to provide tuning rules for high performance applications. This limitation can be overcome incorporating uncertainty models in the linear design, for which several powerful techniques are available in the control literature. However, to the best of our knowledge, these techniques have not yet been used in fuel cell applications. On the other hand, practitioners propose the use of simple proportional or PI controllers around the compressor flow, which ensures good performance in most applications. The latter fact suggests that there is an underlying property of the fuel cell that makes it "easy" to control.

Our objective in this section is twofold: to provide the theoretical justification to the scheme used in practice – revealing its stabilization mechanism – and to give simple tuning rules for high- performance applications. These objectives are attained invoking the property of passivity. As shown in Chapter 2, passive systems can be controlled with simple PI loops whose gains, moreover, can take arbitrary positive values. This last feature is essential to design "tight," high performance and robust, controllers. Robustness is further enhanced if the passivity property is established from the analytic, instead of the numerical, model. In other words, if it holds for all admissible equilibrium points. Our main contribution is to prove that the map from the compressor voltage to the flow is such that, the linearization, around *any* admissible equilibrium, of the third order nonlinear model is *strictly passive*. As a consequence, closing the loop with a PI controller yields – for *a*ll tuning gains and all equilibrium points – a robust (locally) asymptotically stable system. Instrumental to establish this result is the use of the monotonicity characteristics of the system, which stem from the basic physical laws.

Passivity properties have also been exploited in Talj et al. (2009), where the fuel cell system is decomposed into subsystems, one of them being *output strictly passive*. Talj et al. (2009) yields a general methodology to tune all type of controllers – eventually nonlinear – with guaranteed local stability, which is established using linearization. The result presented in this section is different, and in some sense, stronger than the aforementioned one.

The mathematical model of the air supply system proposed in Talj et al. (2009, 2010) and the control problem formulation are presented in Section 3.2.1. Some remarks on the limitations of numerical controller designs and the role of passivity are presented in Section 3.2.2. The linearization of the model and some useful structural properties of it are then given in Section 3.2.3. Section 3.2.4 contains the main result of the section, namely, the proof of passivity of the map. The stability analysis of a PI controller and some simulation results are presented in Sections 3.2.5 and 3.2.6, respectively. We wrap up the section with some concluding remarks and future work in Section 3.2.7.

Further details of the material reported in this chapter may be found in Talj et al. (2011).

3.2.1 Control Problem Formulation

The dynamical equations of the reduced third order model, which suitably captures the behavior of the system are[6]

$$\begin{aligned}
\dot{x}_1 &= -b_1 x_1 + b_2 x_2 + b_3 - b_4 \xi,\\
\dot{x}_2 &= \psi(x_2)\left(\frac{h(x_2,x_3)}{c_{16}} + x_1 - x_2\right),\\
\dot{x}_3 &= -c_9 x_3 - \frac{c_{10}}{x_3}\varphi(x_2)h(x_2,x_3) + c_{13}u
\end{aligned} \tag{3.17}$$

with

$$\begin{aligned}
\varphi(x_2) &= \left(\frac{x_2}{c_{11}}\right)^{c_{12}} - 1,\\
\psi(x_2) &= c_{14}c_{16}\left[1 + c_{15}\varphi(x_2)\right],
\end{aligned}$$

6 The interested reader is referred to Talj et al. (2009, 2010) for further details on the model.

where x_1, x_2 and x_3 are, respectively, the air pressure inside the cathode, the air pressure in the supply manifold (between the compressor and the fuel cell cathode input), and the moto-compressor angular speed. Furthermore, ξ is the fuel cell current, which is a measurable disturbance input, and u is the voltage applied on the compressor motor that is the control input. In view of the difference in time scales between the electrical and the mechanical dynamics, ξ is assumed *constant*. The dynamical equation of x_3 is calculated using the mechanical equation of a simplified DC motor.

All the constants b_i and c_i – that are functions of the physical parameters of the system – are positive. The static function of the compressor, $h(x_2, x_3)$, has the shape shown in Figure 3.2. We draw the reader's attention to the monotonic behavior of the graph, that is essential for the development of our results.

The vector of measurable outputs is

$$y = \begin{bmatrix} y_1 \\ y_2 \\ y_3 \end{bmatrix} = \begin{bmatrix} V(x_1, x_2) \\ x_2 \\ h(x_2, x_3) \end{bmatrix}, \tag{3.18}$$

where y_1 is the fuel cell voltage (also known as polarization characteristic) and y_3 is the compressor flow. See Pukrushpan et al. (2004) for the analytic expressions of the static functions $V(x_1, x_2)$ and $h(x_2, x_3)$, which are omitted here because they are very cumbersome and, for the purposes of our study, only their monotonicity properties are exploited. Note that the moto-

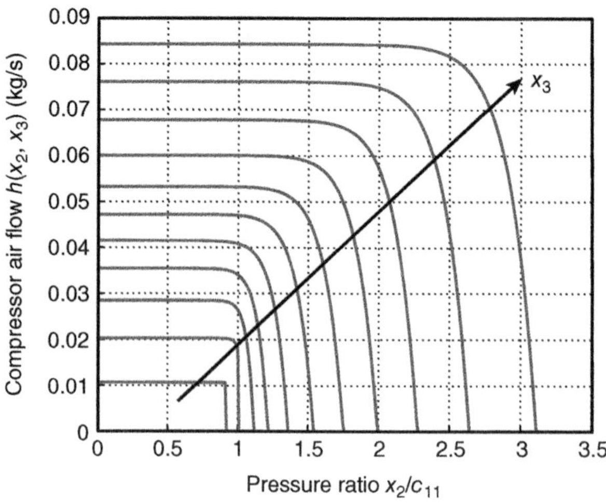

Figure 3.2 Compressor map: the curves are parameterized by x_3, and grow when it increases.

compressor angular speed x_3 is not measurable. In fact, the driving motor and the compressor are both integrated in a closed volume that, due to mechanical constraints, does not include a speed sensor. Speed can, in principle, be recovered from the measurements y_2 and y_3, inverting the function $h(x_2, x_3) = y_3$. But this function is highly nonlinear in x_3, complicating its inversion.

The performance variables for the fuel-cell system are

$$z = \begin{bmatrix} z_1 \\ z_2 \end{bmatrix} = \begin{bmatrix} y_1\xi - c_{21}u(u - c_{22}x_3) \\ \frac{c_{23}}{c_{24}\xi}(x_2 - x_1) \end{bmatrix}, \tag{3.19}$$

where z_1 is the net power delivered to the load and z_2 is the oxygen excess ratio, known also as stoichiometry coefficient. As already mentioned, the control objective is to design a controller that will regulate z_2 around a desired constant value, typically taken to be $z_2^\star = 2$.

The pressures x_1 and x_2 are positive. Furthermore, the moto-compressor is unidirectional; hence, the speed x_3, the flow y_3 and the feeding voltage u are also positive. Finally, the fuel cell is an irreversible device, hence, its current ξ, voltage y_1 and the net power z_1 delivered by the fuel cell are always positive. Note that the stoichiometry z_2 represents a flow ratio and is also positive. Consequently, all the variables of the system belong to the positive orthant.

As indicated above, the control objective is to drive z_2 to a constant desired value. The proposition below shows that this objective can be recast in terms of *stabilization of an equilibrium point*, that we denote $x^\star \in \mathbb{R}^3_+$. Furthermore, it also proves that regulating x_3 to its desired value drives x_1 and x_2 to their equilibrium value.

Proposition 3.2: *Consider the system* (3.17), (3.19) *with ξ fixed to a constant value.*

(i) The equilibrium points $x_1^\star, x_2^\star$ are uniquely defined by $x_3^\star$.
(ii) If $x = x^\star$, then $z_2 = z_2^\star$. Furthermore, for all $z_2^\star$, there exists an $x^\star$, uniquely defined.

Proof. To streamline the proof of the proposition the following observations are in order. First, note that, given an equilibrium $x^\star$, the (constant) control that assigns this equilibrium, say $u^\star$, is univocally defined by the third equation in (3.17) as

$$u^\star := \frac{1}{c_{13}} \left[c_9 x_3^\star + \frac{c_{10}}{x_3^\star}\varphi(x_2^\star)h_3(x_2^\star, x_3^\star) \right]. \tag{3.20}$$

Hence, our attention is concentrated on the first two of equations (3.17). Define the parameterized mapping $r^{\xi} : \mathbb{R}^3_+ \to \mathbb{R}^3$ as

$$r^{\xi}(x) := \begin{bmatrix} -b_1x_1 + b_2x_2 + b_3 - b_4\xi \\ h(x_2,x_3) + c_{16}(x_1 - x_2) \\ x_2 - x_1 - \frac{c_{24}}{c_{23}}\xi z_2^{\star} \end{bmatrix}. \tag{3.21}$$

It is clear that the set $\{x \in \mathbb{R}^3_+ \mid r^{\xi}(x) = 0\}$ identifies the equilibrium points $x^{\star}$ such that $z_2 = z_2^{\star}$. Now, $r^{\xi}(x) = 0$ is a set of three nonlinear algebraic equations in three unknowns. We use the first two equations to prove the first claim. Then, the third equation is solved to find a suitable $x_3^{\star}$.

Proof of (i). The first claim is established showing that the equilibria of the system admit the following parametrization:

$$\begin{aligned} x_1^{\star} &= \frac{1}{b_1}\left[b_2\zeta(x_3^{\star}) + b_3 - b_4\xi\right], \\ x_2^{\star} &= \zeta(x_3^{\star}), \end{aligned} \tag{3.22}$$

where $\zeta : \mathbb{R}_+ \to \mathbb{R}_+$ is a strictly increasing function. Now, fix $x_3^{\star} \in \mathbb{R}_+$. Setting $r_1^{\xi}(x^{\star}) = r_2^{\xi}(x^{\star}) = 0$ get

$$\begin{aligned} x_1^{\star} &= \frac{1}{b_1}\left[b_2x_2^{\star} + b_3 - b_4\xi\right], \\ x_2^{\star} - x_1^{\star} &= \frac{1}{c_{16}}h(x_2^{\star},x_3^{\star}). \end{aligned}$$

Then

$$h(x_2^{\star},x_3^{\star}) = \frac{c_{16}}{b_1}\left[(b_1 - b_2)x_2^{\star} - (b_3 - b_4\xi)\right].$$

In Talj et al. (2009) it is shown that the term $b_1 - b_2$ is positive. Hence, the equilibrium point corresponds to the intersection between the curve $h(x_2,x_3^{\star})$ and a straight line with positive slope. Since $h(x_2,x_3)$ is strictly increasing in x_3 it is clear that there exists a strictly increasing function $\zeta : \mathbb{R}_+ \to \mathbb{R}_+$ satisfying

$$h(\zeta(x_3^{\star}),x_3^{\star}) = c_{16}\left(1 - \frac{b_2}{b_1}\right)\zeta(x_3^{\star}) - \frac{c_{16}}{b_1}(b_3 - b_4\xi), \tag{3.23}$$

completing the proof.

Proof of (ii). In view of the parametrization (3.22), to prove this claim it is enough to show that there exists $x_3^{\star}$ such that $x_3 = x_3^{\star}$ implies $z_2 = z_2^{\star}$. Setting $r_3^{\xi}(x^{\star}) = 0$ and using (3.22) we get

$$\zeta(x_3^{\star}) = \frac{1}{b_1 - b_2}\left(b_1\frac{c_{24}}{c_{23}}\xi z_2^{\star} + b_3 - b_4\xi\right).$$

Now, $\zeta(x_3)$ is a strictly increasing function; therefore, it is one-to-one and admits a left inverse, say $\zeta_L : \mathbb{R}_+ \to \mathbb{R}_+$, such that $\zeta_L(\zeta(x_3)) = x_3$. The proof is completed selecting

$$x_3^\star = \zeta_L \left[\frac{1}{b_1 - b_2} \left(b_1 \frac{c_{24}}{c_{23}} \xi z_2^\star + b_3 - b_4 \xi \right) \right],$$

that, together with (3.22), defines the desired equilibrium. □

3.2.2 Limitations of Current Controllers and the Role of Passivity

In view of the complexity of the nonlinear model – even the reduced third order system (3.17) – most of the controllers for this application are designed based on its linear approximation.[7] Obviously, the analytic model of the linearized dynamics still preserves some of the structural properties of the physical system. Unfortunately, this is lost when *numerical values* are inserted to obtain the model for a *given* equilibrium point. This limits the validity of the analysis – in particular, the predicted stability margins of the controller – to a neighborhood of that particular point. Using numerical *ranges* and designing controllers that are robust to parameter uncertainty, as done in Talj et al. (2009), partially alleviates this problem. As a consequence of losing the physical information, the resulting numerical designs may be conservative, and fail to provide tuning rules for high-performance applications. This significantly complicates the commissioning of the controller, which becomes a tedious and expensive trial-and-error procedure.

As explained above, to overcome the conservativeness problem, we invoke the property of passivity. Our objective is to look for a passive output that should, additionally, be measurable and detectable. In this respect, it is interesting to recall the main result of (Talj et al., 2009), which proves that the nonlinear system (3.17) defines an *output strictly passive* map $u \to x_3$.[8] Unfortunately, the coordinate x_3 is not measurable nor does it define a detectable output. Moreover, this result does not even prove that the linearization of the system is passive. Indeed, it has been shown in Nijmeijer et al. (1992) – see also van der Schaft (2016) – that passivity of the nonlinear system implies (local) passivity of its linearization under some conditions on the storage function, in particular, that it has a minimum at the equilibrium. Unfortunately, the storage function in Talj et al. (2009) is simply x_3^2, which clearly does not satisfy this condition.

7 Piece-wise approximations, used for model predictive controllers, are an alternative that carries the heavy prize of a high computational burden and lack of physical insight.
8 In Talj et al. (2009) the statement is made only for the dynamics of x_3, looking at x_2 as an arbitrary, external signal. It is clear that this is also true for the whole dynamics.

The task of defining an output signal that satisfies all the requirements for the nonlinear model (3.17) seems daunting. Instead, we will content ourselves with the establishment of such a result for the linearized dynamics in Section 3.2.4. Interestingly, it will be shown that the output that is most often used in applications satisfies these properties.

The most common procedure to control the air subsystem (Pukrushpan et al., 2004) is to close the loop with a PI controller around the compressor air flow error $y_3 - h^\star$ – recall that y_3, defined in (3.18), is measurable. Although this configuration ensures good performance in applications, to the best of our knowledge, no rigorous theoretical analysis has been carried out to prove it. A notable exception is (Talj et al., 2009) where, linearizing *only part* of the dynamics, a procedure to tune the gains of a cascaded controller configuration – ensuring stability of the system – is proposed. Unfortunately, consistent with the discussion above, the admissible gain ranges predicted by the theory turned out to be extremely conservative, yielding below par performance (Talj et al., 2010). In the following sections, the remarkable property of passivity of the linearization of the map $u \to (y_3 - h^\star)$ is established.

3.2.3 Model Linearization and Useful Properties

Proposition 3.3: *The linearization of system* (3.17) *around any equilibrium point $x^\star$ is given by*

$$\dot{\eta} = A\eta + b\tilde{u},$$

where $\tilde{u} = u - u^\star$,

$$A = \begin{bmatrix} -b_1 & b_2 & 0 \\ \psi^\star & -\psi^\star\left(1 - \frac{\hbar_2}{c_{16}}\right) & \frac{\psi^\star \hbar_3}{c_{16}} \\ 0 & -\frac{c_{10}}{x_3^\star}\left(\varphi^\star \hbar_2 + \varphi_1 h^\star\right) & -\left[c_9 + c_{10}\frac{\varphi^\star}{x_3^\star}\left(\hbar_3 - \frac{h^\star}{x_3^\star}\right)\right] \end{bmatrix},$$

$$b = \begin{bmatrix} 0 \\ 0 \\ c_{13} \end{bmatrix}, \tag{3.24}$$

and we have defined the constants

$$\varphi^\star := \varphi(x_2^\star), \quad \psi^\star := \psi(x_2^\star), \quad h^\star := h(x_2^\star, x_3^\star),$$

$$\varphi_1 := \frac{\partial \varphi}{\partial x_2}(x_2^\star), \quad \hbar_2 := \frac{\partial h}{\partial x_2}(x_2^\star, x_3^\star), \quad \hbar_3 := \frac{\partial h}{\partial x_3}(x_2^\star, x_3^\star).$$

Proof. The proof is easily established writing the dynamics of the error system as

$$\dot{\eta}_1 = -b_1\eta_1 + b_2\eta_2,$$
$$\dot{\eta}_2 = \psi(x_2)\left(\frac{1}{c_{16}}[h(x_2,x_3) - h^\star] + \eta_1 - \eta_2\right),$$
$$\dot{\eta}_3 = -c_9\eta_3 - c_{10}\left[\frac{1}{x_3}\varphi(x_2)h(x_2,x_3) - \frac{\varphi^\star h^\star}{x_3^\star}\right] + c_{13}\tilde{u}$$

computing the Taylor series expansion and retaining the first terms. □

Although the system (3.17) is a highly complicated set of nonlinear equations, there are several useful properties, which stem from the physical laws, that are instrumental to solve our task. It is shown in Talj et al. (2009), that

$$b_1 > b_2, \tag{3.25}$$
$$\varphi(x_2) > 0,\ \forall x_2 \in \mathbb{R}_+, \tag{3.26}$$
$$\psi(x_2) > 0,\ \forall x_2 \in \mathbb{R}_+. \tag{3.27}$$

Using these properties, and the monotonicity of some of the functions describing the dynamics, it is possible to identify some (sign) properties of the elements of the matrix A.

Proposition 3.4: *The constants appearing in the matrix A verify the following inequalities:*

(C1)

$$\left[c_9 + c_{10}\frac{\varphi^\star}{x_3^\star}\left(\hbar_3 - \frac{h^\star}{x_3^\star}\right)\right] > 0.$$

(C2)

$$\varphi^\star \hbar_2 + \varphi_1 h^\star > 0.$$

(C3)

$$\hbar_2 < 0,\ \hbar_3 > 0.$$

Proof. **(C1)** The function $f(x_3) := \frac{1}{x_3}h(x_2^\star, x_3)$ is strictly increasing for all x_3 in the operating domain. So

$$\frac{df}{dx_3}(x_3) = \frac{1}{x_3^2}\left[\frac{\partial h(x_2^\star, x_3)}{\partial x_3}x_3 - h(x_2^\star, x_3)\right] > 0,$$

in particular, evaluated at the equilibrium, the derivative becomes

$$\frac{df}{dx_3}(x_3^\star) = \frac{1}{x_3^\star}\left(\hbar_3 - \frac{h^\star}{x_3^\star}\right) > 0\,,$$

which yields the desired result, given that $c_9, c_{10}, \varphi^\star > 0$.

(C2) The function $f_2(x_2, x_3) := \varphi(x_2)h(x_2, x_3)$ is strictly increasing with respect to x_2 in the operating domain of the fuel cell. So

$$\frac{\partial f_2}{\partial x_2}(x_2, x_3) = \frac{\partial \varphi}{\partial x_2}(x_2)h(x_2, x_3) + \frac{\partial h}{\partial x_2}(x_2, x_3)\varphi(x_2) > 0,$$

which evaluated at the equilibrium proves the claim.

(C3) From Figure 3.2 it is clear that the function $h(x_2, x_3)$ is strictly decreasing with respect to x_2 and strictly increasing with respect to x_3 – see also the analytic expression in Talj et al. (2009). So

$$\frac{\partial h}{\partial x_2}(x_2, x_3) < 0, \quad \frac{\partial h}{\partial x_3}(x_2, x_3) > 0,$$

which evaluated at the equilibrium completes the proof. □

3.2.4 Main Result

Proposition 3.5: *Consider the system* (3.17) *and an equilibrium point* $x^\star$ *with the output*

$$\tilde{y}_3 := h(x_2, x_3) - h(x_2^\star, x_3^\star). \tag{3.28}$$

(P1) *The linearization of* (3.17), (3.28) *is*

$$\dot{\eta} = A\eta + b\tilde{u}, \quad \tilde{y}_3^\ell = c^\top \eta, \tag{3.29}$$

where (A, b) *are given in* (3.24) *and*

$$c := \mathrm{col}(0, \hbar_2, \hbar_3). \tag{3.30}$$

(P2) *The map* $\tilde{u} \to \tilde{y}_3^\ell$ *is strictly passive. More precisely, along the trajectories of the linearized system*

$$\int_0^t \tilde{u}(\tau)\tilde{y}_3^\ell(\tau)d\tau \geq \alpha \int_0^t |\eta(\tau)|^2 d\tau + \beta$$

for all $t \geq 0$, *all* $\tilde{u}(t)$ *(such that the integrals are well defined) and for some* $\alpha \in \mathbb{R}_+, \beta \in \mathbb{R}$.

Proof. **(P1)** First, note that $\tilde{y}_3$ is zero at the equilibrium. Thus, the linearization of the output, called $\tilde{y}_3^\ell$, is

$$\begin{aligned}\tilde{y}_3^\ell &= \frac{\partial h}{\partial x_2}(x_2^\star, x_3^\star)\eta_2 + \frac{\partial h}{\partial x_3}(x_2^\star, x_3^\star)\eta_3 \\ &= \hbar_2\eta_2 + \hbar_3\eta_3.\end{aligned} \tag{3.31}$$

(P2) The proof of strict passivity is established showing that the transfer function

$$H(s) = c^{\top}(sI_3 - A)^{-1}b, \tag{3.32}$$

is SPR. As stated in Khalil (2002), SPR is tantamount to verifying

(R1) $\Re\left[H(j\omega)\right] > 0,\ \forall\omega \in \mathbb{R}$,
(R2) $\lim_{\omega\to\infty}\left\{\omega^2\Re\left[H(j\omega)\right]\right\} > 0$.

Given the matrices A, b and c in Eqs. (3.24) and (3.30), and after some calculation, one can prove that

$$H(s) = c_{13}\hbar_3.\frac{s^2 + n_1 s + n_0}{s^3 + d_2 s^2 + d_1 s + d_0}, \tag{3.33}$$

where

$$n_1 = b_1 + \psi^\star, \quad n_0 = (b_1 - b_2)\psi^\star,$$

and

$$d_2 = b_1 + \psi^\star\left(1 - \frac{\hbar_2}{c_{16}}\right) + \left[c_9 + c_{10}\frac{\varphi^\star}{x_3^\star}\left(\hbar_3 - \frac{h^\star}{x_3^\star}\right)\right],$$

$$d_1 = b_1\psi^\star\left(1 - \frac{\hbar_2}{c_{16}}\right) - b_2\psi^\star + \left[b_1 + \psi^\star\left(1 - \frac{\hbar_2}{c_{16}}\right)\right]$$
$$\times\left[c_9 + c_{10}\frac{\varphi^\star}{x_3^\star}\left(\hbar_3 - \frac{h^\star}{x_3^\star}\right)\right] + \frac{\psi^\star\hbar_3}{c_{16}}\frac{c_{10}}{x_3^\star}\left(\varphi^\star\hbar_2 + \varphi_1 h^\star\right),$$

$$d_0 = \left[b_1\psi^\star\left(1 - \frac{\hbar_2}{c_{16}}\right) - b_2\psi^\star\right]\left[c_9 + c_{10}\frac{\varphi^\star}{x_3^\star}\left(\hbar_3 - \frac{h^\star}{x_3^\star}\right)\right]$$
$$+ b_1\frac{\psi^\star\hbar_3}{c_{16}}\frac{c_{10}}{x_3^\star}\left(\varphi^\star\hbar_2 + \varphi_1 h^\star\right).$$

Setting $s = j\omega$, the frequency response function becomes

$$H(j\omega) = c_{13}\hbar_3.\frac{R_N + jI_N}{R_D + jI_D},$$

where R_N, I_N, R_D and I_D are the real and imaginary parts of the numerator and the denominator, given by

$$R_N = n_0 - \omega^2, \quad I_N = n_1\omega,$$
$$R_D = d_0 - d_2\omega^2, \quad I_D = \omega(d_1 - \omega^2). \tag{3.34}$$

Hence,

$$\Re\left[H(j\omega)\right] = c_{13}\hbar_3\frac{R_N R_D + I_N I_D}{R_D^2 + I_D^2}. \tag{3.35}$$

Since $c_{13}\hbar_3 > 0$ and $R_D^2 + I_D^2 > 0$, $\Re\left[H(j\omega)\right]$ is positive if and only if $R_N R_D + I_N I_D > 0$. Now,

$$R_N R_D + I_N I_D = \omega^4(d_2 - n_1) + \omega^2(n_1 d_1 - d_0 - n_0 d_2) + n_0 d_0. \tag{3.36}$$

From the monotonicity characteristics of the system and the expressions of n_i and d_i mentioned above, it is clear that

$$d_2 - n_1 = -\psi^\star \frac{\hbar_2}{c_{16}} + \left[c_9 + c_{10}\frac{\varphi^\star}{x_3^\star}\left(\hbar_3 - \frac{h^\star}{x_3^\star}\right)\right] > 0\,, \quad n_0 d_0 > 0. \tag{3.37}$$

Furthermore,

$$\begin{aligned} n_1 d_1 - d_0 - n_0 d_2 = & \left[b_1^2 + (\psi^\star)^2\left(1 - \frac{\hbar_2}{c_{16}}\right) + 2b_2\psi^\star\right] \\ & \times \left[c_9 + c_{10}\frac{\varphi^\star}{x_3^\star}\left(\hbar_3 - \frac{h^\star}{x_3^\star}\right)\right] \\ & + (\psi^\star)^2 \frac{\hbar_3}{c_{16}}\frac{c_{10}}{x_3^\star}\left(\varphi^\star \hbar_2 + \varphi_1 h^\star\right) \\ & - \psi^\star \frac{\hbar_2}{c_{16}}(b_1^\star + b_2\psi^\star) > 0. \end{aligned} \tag{3.38}$$

The inequalities (3.37) and (3.38) show that the term $(R_N R_D + I_N I_D)$ in Eq. (3.36) is positive for all $\omega \in \mathbb{R}$, proving**(R1)**.

Furthermore, given that

$$\begin{aligned} R_D^2 + I_D^2 &= \left(d_0 - d_2\omega^2\right)^2 + \left[\omega(d_1 - \omega^2)\right]^2 \\ &= \omega^6 + (d_2^2 - 2d_1)\omega^4 + (d_1^2 - 2d_0 d_2)\omega^2 + d_0^2, \end{aligned} \tag{3.39}$$

and considering (3.36), we find that

$$\lim_{\omega\to\infty} \left\{\omega^2 \Re\left[H(j\omega)\right]\right\} = c_{13}\hbar_3\,(d_2 - n_1) > 0,$$

where the inequality follows from $c_{13}\hbar_3 > 0$ and $d_2 - n_1 > 0$. □

3.2.5 An Asymptotically Stable PI-PBC

In this section we prove that, in view of strict passivity of the linearized model, the nonlinear system in closed-loop with a PI controller has an asymptotically stable equilibrium at $x^\star$. Although the same result can be established for a simple proportional controller,[9] $\tilde{u} = -k_p\tilde{y}_3$, the addition of the integral action – besides the well-known constant disturbance rejection

9 Actually, for any linear controller that ensures asymptotic stability of the closed-loop.

feature – has the advantage that the controller can be implemented without the knowledge of $u^\star$, which depends on uncertain model parameters.

Proposition 3.6: *Given a fuel cell current ξ and a desired value for the stoichiometry $z_2 = z_2^\star$, let $x^\star$ be the corresponding equilibrium. Consider the system* (3.17) *in closed loop with the PI controller*

$$\begin{aligned} \dot{x}_c &= \tilde{y}_3, \\ u &= -k_p\tilde{y}_3 - k_i x_c, \end{aligned} \tag{3.40}$$

where $\tilde{y}_3$ is given in (3.28). *The equilibrium $(x^\star, -\frac{u^\star}{k_i})$, with $u^\star$ given in* (3.20), *is asymptotically stable, for any $k_i, k_p \in \mathbb{R}_+$. Moreover, $\lim_{t\to\infty} z_2(t) = z_2^\star$.*

Proof. The proof is established invoking Lyapunov's indirect method. That is, we prove that the linearization of the closed-loop system, around the equilibrium $x^\star$, is asymptotically stable, which implies that $x^\star$ is a (locally) asymptotically stable equilibrium of the nonlinear system. The proof that $\lim_{t\to\infty} z_2(t) = z_2^\star$ then follows from claim (ii) in Proposition 3.2.

The linearization of the closed-loop is given by (3.29) together with

$$\begin{aligned} \dot{\tilde{x}}_c &= \tilde{y}_3^\ell, \\ \tilde{u} &= -k_p\tilde{y}_3^\ell - k_i\tilde{x}_c, \end{aligned} \tag{3.41}$$

where we have defined $\tilde{x}_c := x_c - x_c^\star$, and used $x_c^\star = -\frac{u^\star}{k_i}$. The linearized closed-loop system then takes the form

$$\begin{bmatrix} \dot{\eta} \\ \dot{\tilde{x}}_c \end{bmatrix} = \mathcal{A}\begin{bmatrix} \eta \\ \tilde{x}_c \end{bmatrix}, \quad \mathcal{A} := \begin{bmatrix} A - k_p bc^\top & -k_i b \\ c^\top & 0 \end{bmatrix}.$$

We show that $\mathcal{A}$ is Hurwitz proving that it satisfies an algebraic Lyapunov equation of the form

$$\mathcal{P}\mathcal{A} + \mathcal{A}^\top\mathcal{P} = -\mathcal{C}^\top\mathcal{C}, \tag{3.42}$$

where $\mathcal{P} > 0$ and the pair $(\mathcal{A}, \mathcal{C})$ is observable.

Since the transfer function $H(s)$ is SPR, Kalman–Yakubovich–Popov's lemma (Khalil, 2002) yields the existence of positive definite matrices $P \in \mathbb{R}^{3\times3}$ and $Q \in \mathbb{R}^{3\times3}$ such that

$$PA + A^\top P = -Q, \quad Pb = c. \tag{3.43}$$

Consider the positive definite matrix $\mathcal{P} = \begin{bmatrix} P & 0 \\ 0 & k_i \end{bmatrix}$, which yields

$$\mathcal{P}\mathcal{A} + \mathcal{A}^\top\mathcal{P} = -\begin{bmatrix} Q + 2k_p Pbb^\top P & 0 \\ 0 & 0 \end{bmatrix},$$

where (3.43) has been used. The matrix $Q + 2k_p Pbb^\top P$ is positive definite, therefore, it admits a factorization of the form

$$Q + 2k_p Pbb^\top P = K^\top K,$$

where $K \in \mathbb{R}^{3\times3}$ is *nonsingular*. It is clear that (3.42) is satisfied with

$$C = \begin{bmatrix} K & \vdots & 0 \end{bmatrix} \in \mathbb{R}^{3\times4}.$$

The observability claim follows from Popov–Belevitch–Hautus test, e.g. showing that there is no eigenvector of $\mathcal{A}$ in the kernel of C. Indeed, for any vector $v \in \mathbb{C}^4$,

$$Cv = 0 \quad \Leftrightarrow \quad K\mathrm{col}(v_1, v_2, v_3) = 0 \quad \Leftrightarrow \quad v = \mathrm{col}(0, 0, 0, v_4),$$

for some $v_4 \in \mathbb{C}$. Hence,

$$\mathcal{A}v = \begin{bmatrix} -k_i v_4 b \\ 0 \end{bmatrix},$$

and, clearly, there is no (eigenvalue) $\lambda \in \mathbb{C}$ such that $\mathcal{A}v = \lambda v$, which completes the proof. □

3.2.6 Simulation Results

To compute the output for the PI controller (3.40), we write

$$x_1^\star = \frac{z_2^\star \frac{c_{24}}{c_{23}} b_2 - b_4}{b_1 - b_2}\xi + \frac{b_3}{b_1 - b_2}, \tag{3.44}$$

$$x_2^\star = \frac{z_2^\star \frac{c_{24}}{c_{23}} b_1 - b_4}{b_1 - b_2}\xi + \frac{b_3}{b_1 - b_2}, \tag{3.45}$$

which are obtained setting $r_1^\xi(x^\star) = r_3^\xi(x^\star) = 0$ in (3.21). The computation of $h(x_2^\star, x_3^\star)$ is carried out from $r_2^\xi(x^\star) = 0$, with $x_1^\star$ and $x_2^\star$ being replaced by the expressions above, yielding

$$h(x_2^\star, x_3^\star) = \frac{1}{c_{16}}(x_2^\star - x_1^\star) = z_2^\star \frac{c_{24}}{c_{23}}\xi.$$

Hence, $\tilde{y}_3 = h(x_2, x_3) - h(x_2^\star, x_3^\star)$ is known, given that the compressor airflow $y_3 = h(x_2, x_3)$ is measured.

Figures 3.3 and 3.4 present the simulation results of the system (3.17) in closed loop with the PI controller (4.93). The gains of the PI are chosen to be $k_p = 3000$, $k_i = 18000$, so that the poles of the closed-loop system, which are the roots of the characteristic polynomial

$$\begin{aligned} D_{cl}(s) = {} & s^4 + (c_{13}\hbar_3 k_p + d_2)s^3 + \left[c_{13}\hbar_3(k_i + n_1 k_p) + d_1\right] s^2 \\ & + \left[c_{13}\hbar_3(n_1 k_i + n_0 k_p) + d_0\right] s + c_{13}\hbar_3 n_0 k_i, \end{aligned}$$

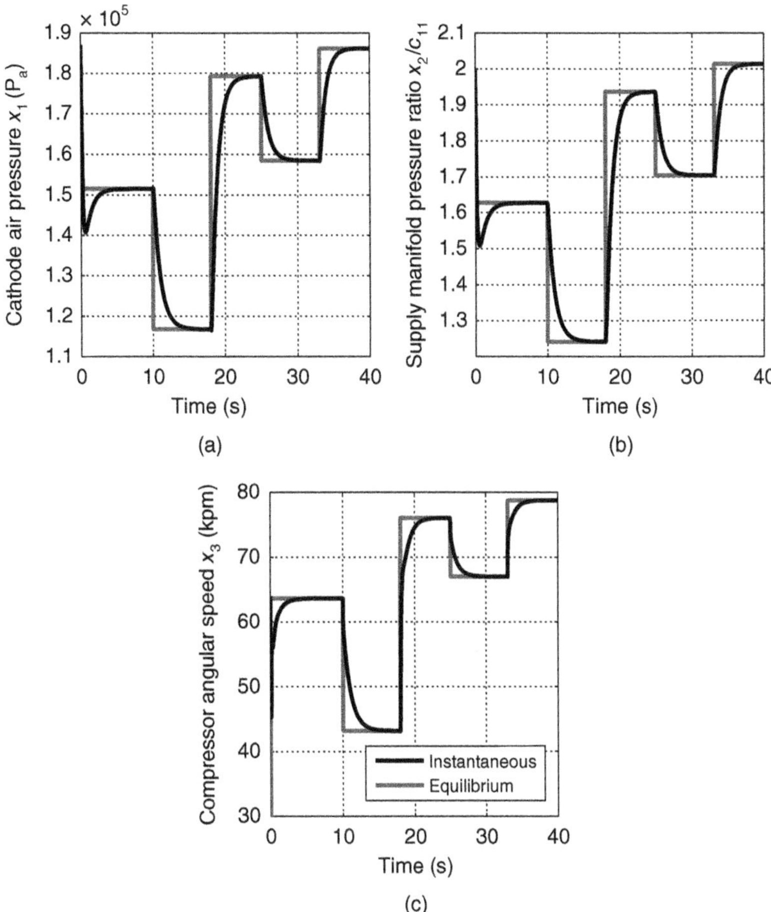

Figure 3.3 Convergence of the three states to the equilibrium point.

are given by

$$p_1 = -1.209, \quad p_2 = -5.663, \quad p_3 = -27.822, \quad p_4 = -54.338,$$

when the fuel cell current is at the nominal value of 191 A. The simulation is performed with the current profile shown in Figure 3.3a. Figure 3.4 shows that the three states x_1, x_2 and x_3, representing, respectively, the air pressure in the cathode, the air pressure in the supply manifold and the rotational speed of the compressor, converge to the equilibrium point corresponding to the desired output $z_2 = z_2^\star$. Figure 3.4b shows the control input, which is the voltage feeding the motor of the compressor. The convergence of the

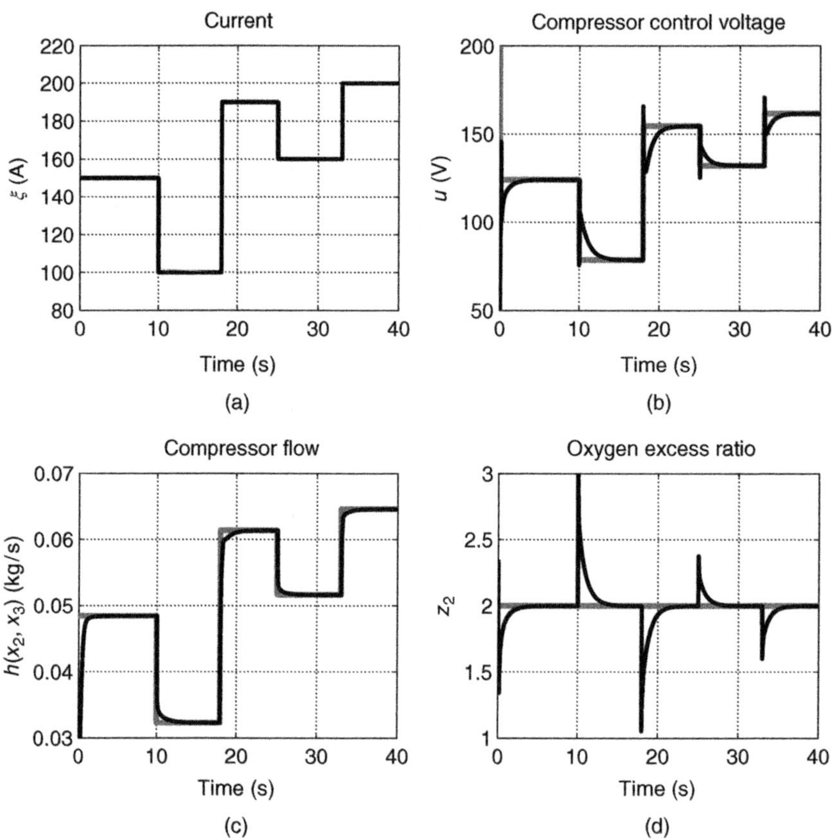

Figure 3.4 Simulation response of the closed-loop system with the PI controller.

compressor airflow to its equilibrium point is shown in Figure 3.4c. Finally, Figure 3.4d shows the convergence of the controlled output z_2 to its desired reference $z_2^\star$, chosen equal to two.

3.2.7 Concluding Remarks and Future Work

In this section, a theoretical justification has been given to the common practice of regulating the air supply system of a PEM fuel cell with a simple PI controller around the compressor flow. Exploiting the monotonic characteristics of some static functions of the fuel cell, it has been verified that the linearized incremental model of the system – at any equilibrium point – is strictly passive. Based on the latter property the stability of the closed-loop system, with a simple PI controller, is proved around any equilibrium point in the operating domain for all $k_p > 0$ and $k_i \geq 0$.

Unfortunately, as shown in Nijmeijer et al. (1992), passivity of the linearized system does not imply (local) passivity of the original nonlinear map. Therefore, no conclusions can be drawn regarding the behavior of the system in closed-loop with a nonlinear controller. However, as proven in van der Schaft (2016), $\mathcal{L}_2$ – stability of the linearized system implies, local, $\mathcal{L}_2$ – stability of the nonlinear system. Since the passivity property that we established is strict, the linearized system is $\mathcal{L}_2$ – stable and small gain theorem arguments can be used to analyze nonlinear controllers. Current research is under way in this direction. It should be pointed out that the passivity property has been established for the linearization at the equilibrium points and not at *any* point in the state space. Some preliminary calculations suggest that the latter, stronger, property is unfortunately not true.

A limitation of the local asymptotic stability result of Proposition 3.6 is that we are unable to estimate the domain of attraction of the stable equilibrium. This requires a Lyapunov analysis of the nonlinear system. Another interesting open question is the potential advantage of using other passive outputs. For instance, it can be shown that the output $h(x_2, x_3) - h(x_2, x_3^\star) = y_3 - h(y_2, x_3^\star)$, which is measurable and detectable, is also a passive output. This output is measurable since the compressor air flow $y_3 = h(x_2, x_3)$ and the air pressure in the supply manifold $y_2 = x_2$ are measurable. Moreover, the analytic function $h(x_2, x_3)$ is known, allowing then the calculation of the term $h(x_2, x_3^\star)$. Simulation results have shown that using the PI with this output, instead of $\tilde{y}_3$, yields some nice transient performance.

Another interesting and important point to be studied is the effect of saturation of the input voltage in fast transients, and its influence on the stability analysis. Saturation is a first-third quadrant static nonlinearity that defines a passive operator. This leads to believe that, as done for mechanical systems, it is possible to carry out this study within our passivity framework.

The result presented in this section provides a first, modest, step toward the development of physically based controller design methodologies, where the mass and energy balance properties of the system are explicitly exploited. This is the essence, and final objective, of the research area generically known as passivity-based control.

Bibliography

F. Blaschke. The principle of field orientation as applied to the new transvector closed-loop control system for rotating-filed machines. *Siemens Review*, 34: 217–220, 1972.

B. K. Bose. *Power Electronics and AC Drives*. Prentice Hall, 1986.

C. Cecati. Experimental investigation of the passivity based induction motor control. In *IEEE International Symposium on Industrial Electronics (ISIE)*, pages 107–112, Guimaraes, Portugal, 1997.

G. Chang, G. Espinosa-Pérez, E. Mendes, and R. Ortega. Tuning rules for the PI gains of field-oriented controller of induction motors. *IEEE Transactions on Industrial Electronics*, 47(3): 592–602, 2000.

C. A. Desoer and M. Vidyasagar. *Feedback Systems: Input-Output Properties*. Academic Press, New York, 2009.

P. A. S. de Wit, R. Ortega, and I. Mareels. Indirect field oriented control of induction motors is robustly globally stable. *Automatica*, 32(10): 1393–1402, 1996.

H. Khalil. *Nonlinear Systems*. Prentice-Hall, Upper Saddle River, NJ, 2002.

K. Kim, R. Ortega, A. Charara, and J. Vilain. Theoretical and experimental comparision of two nonlinear controllers for current-fed induction motors. *IEEE Transactions on Control Systems Technology*, 5(3): 1–11, 1997.

W. Leonhard. *Control of Electrical Drives*. Springer-Verlag, Berlin, 1985.

R. Marino, P. Tomei, and C. M. Verrelli. *Induction Motor Control Design*. Springer Science & Business Media, 2010.

P. J. Nicklasson, R. Ortega, and G. Espinosa-Pérez. Passivity-based control of a class of blondel-park transformable machines. *IEEE Transactions on Automatic Control*, 42(5): 629–647, 1997.

H. Nijmeijer, R. Ortega, A. C. Ruiz, and A. J. van der Schaft. On passive systems: from linearity to nonlinearity. In *IFAC Symposium on Nonlinear Control Systems*, pages 214–219, Bordeaux, France, 1992.

R. Ortega and D. Taoutaou. Indirect field oriented speed regulation of induction motors is globally stable. *IEEE Transactions on Industrial Electronics*, 43(2): 340–341, 1996a.

R. Ortega and D. Taoutaou. On discrete-time control of current-fed induction motors. *Systems and Control Letters*, 28(3), 1996a.

R. Ortega, J. A. Loría, P. J. Nicklasson, and H. Sira-Ramírez. *Passivity-Based Control of Euler-Lagrange Systems*. Springer-Verlag, 1998.

R. Ortega, P. J. Nicklasson, and G. Espinosa-Pérez. On speed control of induction motors. *Automatica*, 32(3): 455–460, 1996.

J. T. Pukrushpan, A. G. Stefanopoulou, and H. Peng. *Control of Fuel Cell Power Systems: Principles, Modeling, Analysis and Feedback Design*. Springer, 2004.

R. Reginatto and A. Bazanella. Robustness of global asymptotic stability in indirect field-oriented control of induction motors. *IEEE Transaction on Automatic Control*, 48(7): 1218–1222, 2003.

R. Talj, D. Hissel, R. Ortega, M. Becherif, and M. Hilairet. Experimental validation of a PEM fuel cell reduced order model and a moto-compressor higher order sliding mode control. *IEEE Transactions on Industrial Electronics*, 57(6): 1906–1913, 2010.

R. Talj, R. Ortega, and A. Astolfi. Passivity and robust PI control of the air supply system of a PEM fuel cell model. *Automatica*, 47: 2554–2561, 2011.

R. Talj, R. Ortega, and M. Hilairet. A controller tuning methodology for the air supply system of a PEM fuel cell system with guaranteed stability properties. *International Journal of Control*, 82(9): 1706–1719, 2009.

D. Taoutaou, R. Puerto, R. Ortega, and L. Loron. A new field-oriented discrete-time controller for current-fed induction motors. *Control Engineering Practice*, 5(2): 209–217, 1997.

A. J. van der Schaft. *L_2-Gain and Passivity Techniques in Nonlinear Control.* Springer-Verlag, Berlin, 3rd edition, 2016.

4

PID-PBC for Nonzero Regulated Output Reference

In this chapter, we provide some answers to the question raised in Chapter 1 regarding the case when the *desired value* for the regulated output – that is, the signal that is fed in the integrator of the PID – is *nonzero*. The most obvious approach to solve this problem, that is the one often used in applications, is to simply wrap the PID around the incremental output and apply the incremental input, that is to propose the PID:

$$\begin{aligned} \dot{x}_c &= y - y^\star, \\ u &= -[K_P(y - y^\star) + K_I x_c + K_D \dot{y}] + u^\star, \end{aligned} \tag{4.1}$$

where $u^\star$ is the value of u at the desired equilibrium – see Appendix B for the definition of assignable equilibria. Because of the presence of the pole at the origin, it is possible to show that if the trajectories converge to an equilibrium, then $y(t) \to y^\star$ as desired. However, the fact that the mapping $u \mapsto y$ is passive, does not guarantee that the trajectories will converge, actually we cannot even ensure that they are bounded. This stems from the fact that, for general nonlinear systems, $u \mapsto y$ passive *does not imply* that $\tilde{u} \mapsto \tilde{y}$ is passive. It is true for LTI systems, a fact that can be proved invoking Kalman–Yakubovich–Popov's lemma (Khalil, 2002). Indeed, if $H(x) = \frac{1}{2}x^\top Px$, with $P \in \mathbb{R}^{n \times n}$ and $P = P^\top > 0$, is a storage function for the original system, then $H(\tilde{x}) = \frac{1}{2}\tilde{x}^\top P\tilde{x}$ is a storage function for the incremental model as well.

In this chapter, we explore two possible solutions to design PID-PBC for the case when $y^\star \neq 0$.

S1 Find a new output, called y_N, which is passive, and moreover, it is such that driving $y_N(t)$ to zero implies that our control objective is satisfied. The definition of the latter may be just that the trajectories are bounded

PID Passivity-Based Control of Nonlinear Systems with Applications, First Edition.
Romeo Ortega, José Guadalupe Romero, Pablo Borja, and Alejandro Donaire.

Published 2021 by John Wiley & Sons, Inc.

and that the following implication is true:

$$\lim_{t\to\infty} y_N(t) = 0 \;\Rightarrow\; \lim_{t\to\infty} y(t) = y^\star,$$

where y the physical output of interest. A more ambitious objective that we consider in this chapter is that

$$\lim_{t\to\infty} y_N(t) = 0 \;\Rightarrow\; \lim_{t\to\infty} |x(t) - x^\star(t)| = 0, \tag{4.2}$$

with $x^\star(t)$ is a desired *trajectory*.

S2 Given that the output of the system is passive, prove that $u \mapsto y$ passive *implies* that $\tilde{u} \mapsto \tilde{y}$ is also passive – a property that is called in the literature *shifted passivity* (van der Schaft, 2016). In this case, we can ensure that the system in closed loop with the PID (4.1) will be $\mathcal{L}_2$-stable.

After presenting the theoretical results pertaining to the two issues discussed above, we illustrate their application to the control of power converters, high-voltage direct-current (HVDC) power systems, wind energy systems, and permanent-magnet synchronous motors (PMSMs).

4.1 PI-PBC for Global Tracking

In this section, we follow the approach proposed in **S1** above to solve the problem of designing PIDs for systems whose control objective is different from regulation to zero of a passive output. As indicated above, we consider the objective of imposing a desired *trajectory* to the state vector and objective that is achieved for a well-defined class of *bilinear* systems, that contains many interesting practical applications.

Bilinear systems are a class of nonlinear systems that describe a broad variety of physical and biological phenomena serving, sometimes, as a natural simplification of more complex nonlinear systems. There is an amount of literature devoted to the study of the intrinsic properties or to stabilization of equilibrium points for these systems. However, to the best of our knowledge, there is no general result for the design of controllers that ensure *global tracking* of (admissible, differentiable) trajectories. The main objective of this section is to provide a theoretical framework for the design of PI-PBCs for global tracking of a class of bilinear systems. Further details of the material reported in this section may be found in Cisneros et al. (2015) and Aranovskiy et al. (2016).

4.1.1 PI Global Tracking Problem

Consider the bilinear system

$$\dot{x} = Ax + d + \sum_{i=1}^{m} u_i B_i x, \tag{4.3}$$

where $x(t), d(t) \in \mathbb{R}^n$ are the state and the (measurable) disturbance vector, respectively, $u(t) \in \mathbb{R}^m$, $m \leq n$, is the control vector, and $A, B_i \in \mathbb{R}^{n\times n}$ are real constant matrices.

Given an admissible, differentiable trajectory, that is a function $x^\star : \mathbb{R}_+ \to \mathbb{R}^n$ verifying

$$\dot{x}^\star = Ax^\star + d + \sum_{i=1}^{m} u_i^\star B_i x^\star \tag{4.4}$$

for some control signal $u^\star : \mathbb{R}_+ \to \mathbb{R}^m$. Find, if possible, an *output* $y_N \in \mathbb{R}^m$ such that the PI controller

$$\begin{aligned} \dot{x}_c &= y_N \\ u &= -K_P y_N - K_I x_c + u^\star \end{aligned} \tag{4.5}$$

ensures that all signals remain bounded and

$$\lim_{t\to\infty} [x(t) - x^\star(t)] = 0, \tag{4.6}$$

for all initial conditions $(x(0), x_c(0)) \in \mathbb{R}^n \times \mathbb{R}^m$.

A set of matrices $\{A, B_i\}$ for which it is possible to solve the aforementioned global tracking problem is identified via the following linear matrix inequality (LMI):

Assumption 4.1: There exists a matrix $P \in \mathbb{R}^{n\times n}$, $P = P^\top > 0$ such that

$$\text{sym}(PA) \leq 0, \tag{4.7}$$

$$\text{sym}(PB_i) = 0. \tag{4.8}$$

To simplify the notation in the sequel, we define the positive semidefinite matrix:

$$Q := -\text{sym}(PA). \tag{4.9}$$

Remark 4.1: Clearly, a particular case of the tracking problem considered here is *regulation* to a constant equilibrium state.

4.1.2 Construction of a Shifted Passive Output

Instrumental to establish the main result of the section is the following lemma:

Lemma 4.1: *Consider the system* (4.3) *verifying the LMI of Assumption 4.1 and an admissible trajectory* $x^\star$*. Define the output function:*

$$y_N := C(x^\star)x, \tag{4.10}$$

where the map $C : \mathbb{R}^n \to \mathbb{R}^{m\times n}$ *is given by*

$$C := \begin{bmatrix} x^{\star\top} B_1^\top \\ \vdots \\ x^{\star\top} B_m^\top \end{bmatrix} P.$$

The operator $\tilde{u} \mapsto y_N$ *is passive with storage function:*

$$V(\tilde{x}) := \frac{1}{2}\tilde{x}^\top P\tilde{x}. \tag{4.11}$$

More precisely,

$$\dot{V} \le \tilde{u}^\top y_N. \tag{4.12}$$

Proof. Combining (4.3) and (4.4) yields

$$\dot{\tilde{x}} = \left(A + \sum_{i=1}^{m} u_i B_i\right)\tilde{x} + \sum_{i=1}^{m} \tilde{u}_i B_i x^\star. \tag{4.13}$$

Now, the time derivative of the storage function (4.11) along the trajectories of (4.13) satisfies

$$\begin{aligned}
\dot{V}(\tilde{x}) &= \tilde{x}^\top P \left[\left(A + \sum_{i=1}^{m} u_i B_i\right)\tilde{x} + \sum_{i=1}^{m} \tilde{u}_i B_i x^\star\right] \\
&= \tilde{x}^\top PA\tilde{x} + \sum_{i=1}^{m} u_i \tilde{x}^\top PB_i\tilde{x} + \sum_{i=1}^{m} \tilde{u}_i \tilde{x}^\top PB_i x^\star \\
&= -\tilde{x}^\top Q\tilde{x} + \sum_{i=1}^{m} \tilde{u}_i \tilde{x}^\top PB_i x^\star \\
&\le \sum_{i=1}^{m} \tilde{u}_i \tilde{x}^\top PB_i x^\star \\
&= \sum_{i=1}^{m} \tilde{u}_i x^\top PB_i x^\star \\
&= y_N^\top \tilde{u},
\end{aligned}$$

where we have used (4.8) of Assumption 4.1 to get the third identity, (4.7) for the first inequality, (4.8) again for the fourth identity and the definition (4.10) for the last one. □

Remark 4.2: A key step for the utilization of the previous result is the derivation of the desired trajectories $x^\star$ and their corresponding control signals $u^\star$, which satisfy (4.4). It is well known that this may prove to be a very complicated task and some approximations may be needed to derive them. Indeed, it is shown in Olm et al. (2011) that even for the simple boost converter this task involves the solution of Abel ordinary differential equation, which is known to be highly sensitive to initial conditions.

Remark 4.3: Notice that, because of the skew-symmetry condition (4.7), the signal $y_N^\star := C(x^\star)x^\star$ is equal to *zero*. Therefore, passivity of the mapping $\tilde{u} \mapsto y_N$ is equivalent to shifted passivity of the system.

4.1.3 A PI Global Tracking Controller

From Lemma 4.1 and Proposition 2.1, the next corollary follows immediately.

Corollary 4.1: *Consider the system* (4.3) *verifying the LMI of Assumption 4.1 and an admissible trajectory* $x^\star$ *which, together with its derivative, is bounded in closed loop with the PI controller* (4.5)*, with the output* y_N *given in* (4.10)*. Then, for all initial conditions* $(x(0), x_c(0)) \in \mathbb{R}^n \times \mathbb{R}^m$*, the trajectories of the closed-loop system are bounded and*

$$\lim_{t\to\infty} y_a(t) = 0, \tag{4.14}$$

where the augmented output $y_a : \mathbb{R}_+ \to \mathbb{R}^{m+n}$ *is defined as*

$$y_a := \begin{bmatrix} C \\ Q \end{bmatrix} \tilde{x}. \tag{4.15}$$

Moreover, if

$$\operatorname{rank} \begin{bmatrix} C \\ Q \end{bmatrix} \geq n, \tag{4.16}$$

then global tracking is achieved, i.e. (4.6) *holds.*

Remark 4.4: Notice that the matrix C depends on the reference trajectory. Therefore, the rank condition (4.16) identifies a class of trajectories for which global tracking of the state is ensured. In the case of regulation to a constant equilibrium, the matrix C is constant, and the passive output y_N is simply a linear combination of the states.

Remark 4.5: Condition (4.16) is sufficient, but not necessary for convergence of the state error. Indeed, as shown in Section A.3, global tracking is guaranteed if y_N is a detectable output for the closed-loop system. That is, if the following implication holds

$$y_N(t) \equiv 0 \implies \lim_{t\to\infty} [x(t) - x^\star(t)] = 0, \tag{4.17}$$

which is strictly weaker than the rank condition (4.16).

4.2 Conditions for Shifted Passivity of General Nonlinear Systems

In this section, we follow the approach proposed in **S2** above – that is, to prove shifted passivity – to solve the problem of designing PIDs for general nonlinear systems whose control objective is different from regulation to zero of a passive output. Further details of the material reported in this section may be found in Jayawardhana et al. (2007).

4.2.1 Shifted Passivity Definition

Given a nonlinear system of the form

$$\begin{aligned} \dot{x} &= f(x) + g\,u, \\ y &= h(x), \end{aligned} \tag{4.18}$$

where $x(t) \in \mathbb{R}^n, u(t), y(t) \in \mathbb{R}^m$, with $m \leq n$, the functions f, h are locally Lipschitz and the matrix $g \in \mathbb{R}^{n\times m}$ is *constant* and has full rank. Fix an equilibrium state $x^\star \in \mathbb{R}^n$, that is,

$$x^\star \in \mathcal{E} := \{x \in \mathbb{R}^n \mid g^\perp f(x) = 0\}, \tag{4.19}$$

where $g^\perp \in \mathbb{R}^{(n-m)\times n}$ is a full-rank left-annihilator of g, i.e. $g^\perp g = 0$ and $\text{rank}\{g^\perp\} = n - m$, and define the constant input and output vectors associated to $x^\star$ as

$$\begin{aligned} u^\star &:= (g^\top g)^{-1} g^\top f(x^\star), \\ y^\star &:= h(x^\star). \end{aligned} \tag{4.20}$$

See Section B.2 for further discussion on assignable equilibria. Define the incremental model:

$$\begin{aligned} \dot{x} &= f(x) + g u^\star + g\tilde{u}, \\ \tilde{y} &= h(x) - h(x^\star). \end{aligned} \tag{4.21}$$

Assume (4.18) defines a passive mapping $u \to y$. Under which conditions the mapping $\tilde{u} \to \tilde{y}$ is also passive? Systems verifying this property are said to be *shifted passive*.[1]

Remark 4.6: Note that shifted passivity is different from the classical *incremental* passivity property (Desoer and Vidyasagar, 2009). In fact, the latter is much more demanding as the word "incremental" refers to two *arbitrary* input–output pairs of the system, whereas in the former only one input–output pair is arbitrary, and the other one is fixed to a constant. Notice that shifted passivity is defined with respect to a given pair $(x^\star, u^\star) \in \mathcal{E}$. If the shifted passivity property holds for all $(x^\star, u^\star) \in \mathcal{E}$, then the (shifted) passivity property becomes independent of the steady-state values $u^\star$ and $x^\star$ (Hines et al., 2011).

Remark 4.7: It is interesting to note that the dissipation obstacle[2] is conspicuous by its absence in shifted passive systems. Indeed, the dissipation obstacle is characterized by the presence of (pervasive) dissipation that makes the supplied power evaluated at the equilibrium nonzero, that is, $(u^\star)^\top y^\star \neq 0$ – this supplied power is zero for shifted passive systems.

4.2.2 Main Results

Proposition 4.1: *Assume*

A.1 *The system* (4.18) *defines a passive mapping* $u \mapsto y$ *with a* convex *twice continuously differentiable storage function* $H : \mathbb{R}^n \to \mathbb{R}_+$.

A.2 *The condition*

$$[f(x) - f(x^\star)]^\top [\nabla H(x) - \nabla H(x^\star)] \leq 0 \tag{4.22}$$

is satisfied.

Then, the mapping $\tilde{u} \mapsto \tilde{y}$, *defined by* (4.21) *is also passive with nonnegative storage function* $H_B : \mathbb{R}^n \to \mathbb{R}_+$,

$$H_B(x) = H(x) - x^\top \nabla H(x^\star) - [H(x^\star) - (x^\star)^\top \nabla H(x^\star)]. \tag{4.23}$$

Proof. We will verify that **A.1** and **A.2** ensure that (4.21) satisfies the necessary and sufficient conditions for passivity of Hill–Moylan's theorem given in Section A.1 (Hill and Moylan, 1980) with the storage function

1 This property was called "Passivity of the incremental model" in Jayawardhana et al. (2007).

2 See Ortega et al. (2001) and Chapter 2.

$H_B(x)$. Namely, the condition of stability (with Lyapunov function $H_B(x)$) of $x^\star$:

$$[f(x) + gu^\star]^\top \nabla H_B(x) \leq 0, \tag{4.24}$$

and the coupling condition between input and the (new) output mappings, that is,

$$h(x) - h(x^\star) = g^\top \nabla H_B(x). \tag{4.25}$$

From (4.23), we have that

$$\nabla H_B(x) = \nabla H(x) - \nabla H(x^\star). \tag{4.26}$$

Now, because of our assumption of constant g, $f(x^\star) + gu^\star = 0$ for all equilibrium points. Replacing the latter in (4.22) and using (4.26) clearly yields (4.24). The second condition, (4.25), follows immediately from (4.26) and the fact that passivity of (4.18) implies

$$h(x) = g^\top \nabla H(x).$$

It only remains to prove that $H_B(x)$ is nonnegative, which will be done showing that $x^\star$ is a minimum point of $H_B(x)$. From (4.26), we obtain $\nabla H_B(x^\star) = 0$. On the other hand, using convexity of $H(x)$, we get

$$\nabla^2 H_B(x) = \nabla^2 H(x) \geq 0. \tag{4.27}$$

This completes the proof. □

For the purposes of stability analysis, it is often of interest to investigate whether the storage function has an *unique* minimum at $x^\star$ and whether the function is proper. As shown in the proof of the proposition, $H_B(x)$ has indeed a minimum at $x^\star$, but this may not be unique. The proposition 4.2 shows that having a strictly convex $H(x)$ is sufficient to ensure that $H_B(x)$ has a unique minimum at $x^\star$ and, furthermore, that it is proper. The proof of this result, being quite technical is omitted, and may be found in Jayawardhana et al. (2007).

Proposition 4.2: *Assume the storage function $H(x)$ is strictly convex. Then, for every $x^\star \in \mathbb{R}^n$, the new storage function $H_B(x)$ defined in (4.23) has a unique global minimum at $x^\star$ and is proper.*

Remark 4.8: The shifted function $H_B(x)$ is closely related to the Bregman divergence in convex analysis (Bregman, 1967), and to the availability function as used in thermodynamics (Alonso and Ydstie, 2001; Keenan, 1951). A by-product of the construction of shifted storage functions is a passivity property which is uniform for a range of steady-state solutions. This is particularly advantageous in flow networks, distribution, and electrical networks where loads/demands are not precisely known (Wen and Arcak, 2004; de Persis et al., 2014; Trip et al., 2016); see also Hines et al. (2011), Bürger et al. (2014), and Simpson-Porco (2019) where the term "equilibrium-independent passivity" has been used to refer to a closely related passivity property.

Remark 4.9: In the LTI case with quadratic storage function, Eq. (4.22) reduces to the stability condition $\tilde{x}^\top PA\tilde{x} \leq 0$, while the new storage function is given by $H_B(\tilde{x}) = \frac{1}{2}\tilde{x}^\top P\tilde{x}$. This appealing downward compatibility makes our result a natural extension, to the nonlinear case, of the well-known property of LTI systems.

Remark 4.10: If (4.18) is a pH-system, we have

$$f(x) = [\mathcal{J}(x) - \mathcal{R}(x)]\nabla\mathcal{H}(x),$$

where $\mathcal{J}(x) = -\mathcal{J}^\top(x)$ and $\mathcal{R}(x) = \mathcal{R}^\top(x) \geq 0$. Condition (4.22) will be then satisfied if $\mathcal{J}$ and $\mathcal{R}$ are *constant* matrices. Stronger results for this particular case of pH systems are given in Section 4.3.

Remark 4.11: The results reported in this section were instrumental to prove that a large class of nonlinear RLC circuits are GAS with PI-PBC (Castaños et al., 2009).

4.3 Conditions for Shifted Passivity of Port-Hamiltonian Systems

In this section, we give conditions for shifted passivity of pH systems. Further details of the material reported in this section may be found in Monshizadeh et al. (2019).

4.3.1 Problems Formulation

In Section 4.2, we showed that pH systems with convex Hamiltonian are also shifted passive provided the input, dissipation and interconnection matrices are all constant. Conditions for shifted passivity of pH systems with state-dependent matrices have been reported in Maschke et al. (2000) and Ferguson et al. (2015). In the former case, quite conservative, integrability conditions, are imposed while the latter ones can be difficult to verify. The main result of this section is to give compact and easily verifiable conditions – i.e. monotonicity of a suitably defined function – to ensure shifted passivity of pH systems with strictly convex Hamiltonian and *state-dependent* dissipation and interconnection matrices. Notably, for the case of affine pH systems with quadratic Hamiltonian, our conditions reduce to negative semidefiniteness of a *constant* matrix. The latter becomes a necessary and sufficient condition for shifted passivity in case the input matrix is nonsingular.

The proposed conditions are exploited to certify local and global stability of *forced* pH systems, i.e., under constant external inputs. An additional contribution in the section is that the proposed conditions provide an estimate of the excess and shortage of passivity that serves as a tool for controller design.

Consider the pH system

$$\dot{x} = [\mathcal{J}(x) - \mathcal{R}(x)]\nabla H(x) + Gu, \tag{4.28a}$$

$$y = G^\top \nabla H(x), \tag{4.28b}$$

with state $x(t) \in \mathrm{R}^n$, input $u(t) \in R^m$, and output $y(t) \in \mathrm{R}^m$. The constant matrix $G \in \mathbb{R}^{n\times m}$ has full column rank, and $H : \mathbb{R}^n \to \mathbb{R}$ is the Hamiltonian of the system.[3] The matrix $\mathcal{J}$ is skew-symmetric, i.e., $\mathcal{J}(x) + \mathcal{J}^\top(x) = 0$, and

$$\mathcal{R}(x) \geq R_{\min}, \ \forall x \in \mathbb{R}^n \tag{4.29}$$

for some constant positive semidefinite matrix $R_{\min}$.

Define the steady-state relation

$$\mathcal{E}_r := \{(x, u) \in \mathbb{R}^n \times \mathbb{R}^m \mid F(x)\nabla H(x) + Gu = 0\},$$

where to simplify the notation we defined the matrix $F(x) := \mathcal{J}(x) - \mathcal{R}(x)$.

Fix $(x^\star, u^\star) \in \mathcal{E}$ and the corresponding output $y^\star := G^\top \nabla H(x^\star)$.

We are interested in providing answers to the following problems:

P1 Find conditions on the pH system (4.28) such that the mapping $(u - u^\star) \to (y - y^\star)$ is passive.

P2 Find conditions under which shifted passivity can be enforced via output-feedback, i.e. the system is output-feedback shifted passifiable.

P3 Use the shifted passivity property to study the stability of the pH system with a constant external input.

P4 Give necessary and sufficient conditions for shifted passivity of pH systems with quadratic dependence on x.

3 The case of state-dependent input matrix $G(x)$ will be discussed later.

4.3.2 Shifted Passivity

Here, we provide conditions under which the pH system (4.28) is shifted passive. Toward this end, we make two assumptions:

Assumption 4.2: The Hamiltonian H is strictly convex.

To state the second assumption, it is convenient to express the matrix $F(x)$ in terms of the coenergy variables, that is, in terms of the signal $s := \nabla H(x)$. Toward this end, we recall the Legendre transform of H as the function:

$$H^{\star}(p) := \max_{x \in \mathbb{R}^n} \{x^{\top} p - H(x)\}, \tag{4.30}$$

where the domain of $H^{\star}$ is the set of all p for which the expression is well-defined (i.e. the maximum is attained), see Rockafellar (2015). A key property of the Legendre transformation that we exploit in the sequel is that[4]

$$\nabla H^{\star}(\nabla H(x)) = x, \ \forall x.$$

Leveraging the last property, the function $F(x)$ can be expressed in terms of coenergy variables as

$$F(x) = F(\nabla H^{\star}(s)) =: \mathcal{F}(s). \tag{4.31}$$

We are in position to state our next assumption:

Assumption 4.3: The mapping $\mathcal{F}$ verifies

$$\nabla(\mathcal{F}(s)\, s^{\star}\ ) + \nabla(\mathcal{F}(s)\, s^{\star}\ )^{\top} - 2R_{\min} \leq 0, \ \forall s \in \mathcal{S}, \tag{4.32}$$

where $\mathcal{S}$ is the range of ∇H.

Notice that Assumption 4.3 is trivially satisfied if the matrices $\mathcal{J}$ and $\mathcal{R}$ in (4.28) are state-independent.

Now, we have the following result, whose proof may be found in Monshizadeh et al. (2019).

Proposition 4.3: *Let Assumptions 4.2 and 4.3 hold. Then, the pH system (4.28) is shifted passive, namely*

$$\dot{H}_B \leq (u - u^{\star})^{\top}(y - y^{\star}) \tag{4.33}$$

is satisfied, with H_B defined in (4.23).

4 In words, this property means that the gradient of the Legendre transform is equal to the inverse of ∇H.

4.3.3 Shifted Passifiability via Output-Feedback

In this section, we consider the case where the condition (4.32) does not hold, which means that the system (4.28) may not be shifted passive, but it can be rendered shifted passive via output feedback.

Proposition 4.4: *Consider the pH system (4.28) verifying Assumption 4.2 and such that*

$$\nabla(\mathcal{F}(s)\, s^\star\,) + \nabla(\mathcal{F}(s)\, s^\star\,)^\top - 2R_{\min} \leq 2\kappa\; GG^\top,$$

for some $\kappa \in \mathbb{R}$. Then, the shifted Hamiltonian (4.23) *satisfies the following dissipation inequality:*

$$\dot{H}_B \leq (u - u^\star)^\top (y - y^\star) + \kappa |y - y^\star|^2.$$

Note that a negative κ proves that the pH system is (output-strictly) shifted passive. On the other hand, a positive κ indicates the shortage of shifted passivity. Notice that the simple proportional controller

$$u = u^\star - K_P(y - y^\star) + v,$$

with $K_P \geq \kappa I_m$, ensures that the closed-loop system is passive from the external input v to output $y - y^\star$.

4.3.4 Stability of the Forced Equilibria

Lyapunov stability of the equilibrium of (4.28) with $u = u^\star$, immediately follows from Proposition 4.3, with the Lyapunov function being the shifted Hamiltonian H_B. Moreover, asymptotic stability follows by imposing the condition that $\dot{H}_B$ is negative definite. As will be shown below, for stability analysis, we can in fact drop the assumption that the input matrix is constant. To this end, consider again the pH system (4.28a), where now $G = G(x)$. Let

$$G(x) = G(\nabla H^\star(s)) =: \mathcal{G}(s).$$

Then, we have the following result, whose proof may be found in Monshizadeh et al. (2019).

Proposition 4.5: *Consider the pH system*

$$\dot{x} = F(x)\nabla H(x) + G(x)u^\star, \tag{4.34}$$

with $(x^\star, u^\star) \in \mathcal{E}$. Then, we have

1. *The equilibrium is asymptotically stable if* $\nabla^2 H(x^\star) > 0$ *and there exists* $\epsilon > 0$ *such that the inequality*

$$\nabla(\mathcal{F}(s)\, s^\star + \mathcal{G}(s)u^\star) + [\nabla(\mathcal{F}(s)\, s^\star + \mathcal{G}(s)u^\star)]^\top - 2R_{\min} \leq -2\epsilon I_n, \tag{4.35}$$

holds at $s = s^\star$.[5]

2. *The equilibrium is globally asymptotically stable if the Hamiltonian H is strongly convex and* (4.35) *holds for all* $s \in \mathrm{R}^n$.

Remark 4.12: The identity matrix in the right hand side of (4.35) can be replaced by a positive semidefinite matrix $C^\top C$, with $C \in \mathrm{R}^{m\times n}$, if the equilibrium is "observable" from the input–output pair $(u^\star, C\nabla H(x^\star))$, namely if

$$\dot{x} = F(x)\nabla H + G(x)u^\star, \quad C\nabla H(x) = C\nabla H(x^\star) \Longrightarrow x = x^\star.$$

4.3.5 Application to Quadratic pH Systems

In this section, we specialize our results to the case where

$$F(x) = F_0 + \sum_{i=1}^{n} F_i x_i, \tag{4.36}$$

with $F_j \in \mathbb{R}^{n\times n}$, $j = 0, \dots, n$, constant and

$$H(x) = \frac{1}{2} x^\top Q x, \tag{4.37}$$

with $Q \in \mathbb{R}^{n\times n}$ being positive definite. We call these systems quadratic pH systems.

In order to satisfy (4.29) and state the global version of our results, we need to assume that $\mathcal{R}(x) = R_0$ for some constant matrix R_0. This is due to the fact that in the affine case the inequality $\mathcal{R}(x) \geq R_{\min}$, for all $x \in \mathbb{R}^n$, implies that the matrix $\mathcal{R}$ is constant. Note that, in this case, $F_0 + F_0^\top = -2R_0 \leq 0$ and $F_j + F_j^\top = 0$ for each $j \geq 1$.

The main result of the subsection is given in the proposition below whose proof may be found in Monshizadeh et al. (2019).

Proposition 4.6: *Consider the quadratic pH system (4.28) with* (4.36) *and* (4.37). *Fix* $(x^\star, u^\star) \in \mathcal{E}$ *and define the* $n \times n$ constant *matrix:*

$$B := \sum_{i=1}^{n} F_i Q\, x^\star\, e_i^\top Q^{-1}, \tag{4.38}$$

5 This means that the Jacobian of $\mathcal{F}(s)\, s^\star + \mathcal{G}(s)u^\star$ in (4.35) has to be evaluated at $s = s^\star$.

with $e_i \in \mathbb{R}^n$ the ith element of the standard basis. Then, we have

$$\dot{H}_B \leq (y - y^\star)^\top (u - u^\star), \tag{4.39}$$

where $H_B(x) := \frac{1}{2}(x - x^\star)^\top Q(x - x^\star)$, if and only if

$$B + B^\top - 2R_0 \leq 0. \tag{4.40}$$

The proposition above can be combined with Proposition 4.3 to give the following simple condition for global asymptotic stability of forced quadratic pH systems.

Corollary 4.2: *Consider the quadratic affine pH system (4.34), (4.36), (4.37), with $G(x) = G_0 + \sum_{i=1}^{n} G_i x_i$, where $G_j \in \mathbb{R}^{n\times m}$, $j = 0, \ldots, n$, are constant. Then the equilibrium $x^\star$ is GAS if*

$$\tilde{B} + \tilde{B}^\top - 2R_0 < 0,$$

where

$$\tilde{B} := B + \sum_{i=1}^{n} G_i u^\star e_i^\top Q^{-1}.$$

4.4 PI-PBC of Power Converters

In this section, we apply the results of Section 4.1 to design PI-PBC for a general class of switched power converters. More precisely, we construct an output y_N such that the implication (4.2), with $x^\star \in \mathbb{R}^n$ a *constant, desired equilibrium*, holds true. Further details of the material reported in this section may be found in Hernandez-Gomez et al. (2010) and Hernandez-Gomez et al. (2012).

4.4.1 Model of the Power Converters

We consider power converters described by the dynamics given in Section D.2. The control objective is to stabilize with a PI controller an equilibrium of the system (D.6). As proposed in Section 4.1, we define an output $y_N \in \mathbb{R}^m$ via (4.10) – that, in the regulation case, is a *linear* combination of the states – such that the mapping $\tilde{u} \mapsto y_N$ is passive[6] and, moreover, the implication (4.2) holds true. As shown in Corollary 4.1, in this case global asymptotic stability of $x^\star$ is achieved with the PI controller (4.5).

6 Recall from Remark 4.3, that this is equivalent to shifted passivity.

Remark 4.13: In the foundational paper (Sanders and Verghese, 1992), it was shown that the nonlinear incremental model of power converters is passive and a stabilizing switching control for the duty ratio was proposed. As indicated in Sanders and Verghese (1992), passivity of the nonlinear incremental model is a particular case of incremental passivity (Desoer and Vidyasagar, 2009) and it is called in that paper "relative passivity." See also Pérez et al. (2004).

4.4.2 Construction of a Shifted Passive Output

Proposition 4.7: *Consider switched power converters described by (D.6). Let $x^\star \in \mathbb{R}^n$ be an admissible equilibrium point, that is, $x^\star$ satisfies*

$$0 = \left(J_0 + \sum_{i=1}^{m} J_i u_i^\star - R \right) \nabla H(x^\star) + \left(G_0 + \sum_{i=1}^{m} G_i u_i^\star \right) E, \tag{4.41}$$

for some $u^\star \in \mathbb{R}^m$. The output $y_N = Cx$, where

$$C := \begin{bmatrix} E^\top G_1^\top - (x^\star)^\top Q J_1 \\ \vdots \\ E^\top G_m^\top - (x^\star)^\top Q J_m \end{bmatrix} Q \in \mathbb{R}^{m \times n} \tag{4.42}$$

is such that the mapping $\tilde{u} \mapsto \tilde{y}_N$ is passive – hence, the system is shifted passive. More precisely, the system verifies

$$\dot{V} \leq \tilde{y}_N^\top \tilde{u},$$

with the (positive definite) storage function $V : \mathbb{R}^n \to \mathbb{R}_+$ given by

$$V(x) = \frac{1}{2} \tilde{x}^\top Q \tilde{x}. \tag{4.43}$$

Proof. To simplify the notation, define the matrix function $G_N : \mathbb{R}^n \to \mathbb{R}^{n \times m}$:

$$G_N(x) := [J_1 Q x + G_1 E \mid \cdots \mid J_m Q x + G_m E], \tag{4.44}$$

which allows to write (D.6) in the compact form

$$\dot{x} = \left(J_0 - R\right) \nabla H(x) + G_0 E + G_N(x) u.$$

Now,

$$\begin{aligned} \dot{x} &= \left(J_0 - R\right) \nabla V(x) + G_N(x)(\tilde{u} + u^\star) - G_N(x^\star) u^\star \\ &= \left(J_0 + \sum_{i=1}^{m} J_i u_i^\star - R \right) \nabla V(x) + G_N(x) \tilde{u}, \end{aligned}$$

where (4.41) and (4.43) have been used to obtain the first identity and

$$[G_N(x) - G_N(x^\star)]u^\star = \sum_{i=1}^{m} J_i u_i^\star Q\tilde{x},$$

which follows from (4.44), is used to get the second one.

Now, we see that $\tilde{y}_N$ may be written as

$$\tilde{y}_N = G_N^\top(x^\star)Q\tilde{x}.$$

On the other hand, it is easy to verify that

$$G_N^\top(x^\star)Q\tilde{x} = G_N^\top(x)Q\tilde{x}.$$

The proof is completed evaluating the derivative of (4.43) along the trajectories of the system, replacing the expression of $V(x)$ and using the nonnegativity of R to get

$$\begin{aligned} \dot{V} &= -\nabla^\top V(x) R \nabla V(x) + \tilde{y}_N^\top \tilde{u} \\ &= -\tilde{x}^\top QRQ\tilde{x} + \tilde{y}_N^\top \tilde{u} \\ &\leq \tilde{y}_N^\top \tilde{u}. \end{aligned} \tag{4.45}$$

□

Remark 4.14: An interesting observation stemming from the skew-symmetry of J_i is that

$$y_N^\star = Cx^\star = \begin{bmatrix} E^\top G_1^\top \\ \vdots \\ E^\top G_m^\top \end{bmatrix} Qx^\star, \tag{4.46}$$

which has two important consequences, first, that $y_N^\star$ is linear in $x^\star$ – a property that is essential for adaptive designs (Hernandez-Gomez et al., 2010, Jayawardhana et al., 2007). Second, that $y_N^\star = 0$ for converters *without* switched external sources.

4.4.3 PI Stabilization

An immediate corollary of passivity of the incremental model is the following result, whose proof is established from a direct application of the passivity theorem.

Proposition 4.8: *Consider a switched power converter described by (D.6) in closed loop with the PI controller:*

$$\begin{aligned} \dot{x}_c &= \tilde{y}_N, \\ u &= -K_P \tilde{y}_N - K_I x_c, \end{aligned} \tag{4.47}$$

with $\tilde{y}_N = C\tilde{x}$ *and* C *given by* (4.70), *and* $K_P = K_P^\top > 0$, $K_I = K_I^\top > 0$ *and* $x^\star, u^\star$ *satisfying* (4.41). *For all initial conditions* $(x(0), z(0)) \in \mathbb{R}^{n+m}$ *the trajectories of the closed-loop system are bounded and such that*

$$\lim_{t\to\infty} y_a(t) = 0,$$

where y_a *is an augmented "output signal" defined as*

$$y_a := \begin{bmatrix} C \\ RQ \end{bmatrix} \tilde{x}. \tag{4.48}$$

Moreover,

$$\lim_{t\to\infty} x(t) = x^\star,$$

if y_a *is detectable, that is, if for any solution* $x(t)$ *of the closed-loop system, the following implication is true:*

$$y_a(t) \equiv 0 \quad \Rightarrow \quad \lim_{t\to\infty} x(t) = x^\star. \tag{4.49}$$

Remark 4.15: The detectability condition (4.49) of Proposition 4.8 may be verified by computing the zero dynamics (associated to the output $\tilde{y}_a$) and then proving that it is asymptotically stable. However, this analysis may be quite involved. In Hernandez-Gomez et al. (2010) verifiable conditions on the matrices J_i and R that ensure detectability are given. Also, invoking condition (4.16) of Corollary 4.1, we see that global asymptotic stability is ensured if $R > 0$ – which is the case if all energy storing elements are lossy.

4.4.4 Application to a Quadratic Boost Converter

In this section, we apply the PI-PBC described above to the quadratic converter shown in Figure 4.1. Its dynamic behavior is described with the averaged model given by the equations:

$$\begin{aligned}
\dot{i}_{L1} &= \frac{1}{L_1}(E - v_{C1}u), \\
\dot{i}_{L2} &= \frac{1}{L_2}(v_{C1} - v_{C2}u), \\
\dot{v}_{C1} &= \frac{1}{C_1}(i_{L1}u - i_{L2}), \\
\dot{v}_{C2} &= \frac{1}{C_2}\left(i_{L2}u - \frac{1}{r_L}v_{C2}\right),
\end{aligned} \tag{4.50}$$

where $i_{L1}(t), i_{L2}(t)$ are the currents in the inductances, $v_{C1}(t), v_{C2}(t)$ are the voltages in the capacitors, L_1, L_2, C_1 and C_2 are the values of inductances and

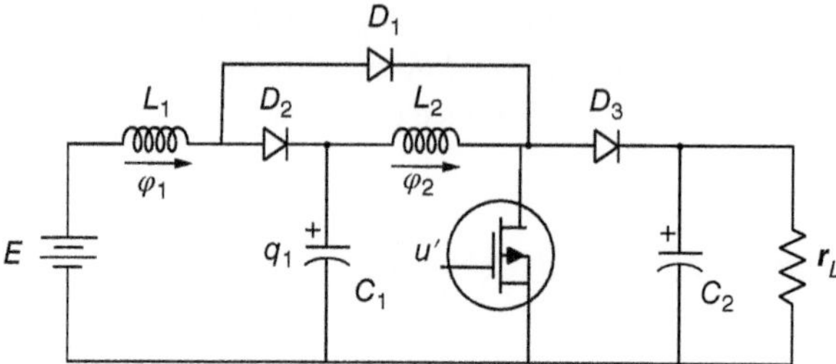

Figure 4.1 Schematic of the quadratic boost converter.

capacitances, respectively, r_L is the load, E the input voltage, and $u = 1 - u'$ is a continuous control equal to the slew rate of the pulse-width modulator (PWM). To simplify the analysis, the inductors are assumed lossless.

To apply the construction of the PI-PBC of Proposition 4.8, the model (4.50) is expressed in the pH form (D.6) with the definitions

$$x = \begin{pmatrix} i_{L1} & i_{L2} & v_{C1} & v_{C2} \end{pmatrix}^\top, \quad B = \begin{pmatrix} \frac{E}{L_1} & 0 & 0 & 0 \end{pmatrix}^\top,$$

$$R = \begin{pmatrix} 0 & 0 & 0 & 0 \\ 0 & 0 & 0 & 0 \\ 0 & 0 & 0 & 0 \\ 0 & 0 & 0 & \frac{1}{r_L C_2^2} \end{pmatrix}, \quad Q = \begin{pmatrix} L_1 & 0 & 0 & 0 \\ 0 & L_2 & 0 & 0 \\ 0 & 0 & C_1 & 0 \\ 0 & 0 & 0 & C_2 \end{pmatrix},$$

$$J_0 = \frac{1}{C_1 L_2} \begin{pmatrix} 0 & 0 & 0 & 0 \\ 0 & 0 & 1 & 0 \\ 0 & -1 & 0 & 0 \\ 0 & 0 & 0 & 0 \end{pmatrix}, \quad J_1 = \begin{pmatrix} 0 & 0 & -\frac{1}{L_1 C_1} & 0 \\ 0 & 0 & 0 & -\frac{1}{L_2 C_2} \\ \frac{1}{L_1 C_1} & 0 & 0 & 0 \\ 0 & \frac{1}{L_2 C_2} & 0 & 0 \end{pmatrix},$$

where $H(x) = \frac{1}{2} x^\top Q x$ is the energy stored in the circuit.

The goal of the control is to regulate the voltage v_{C2} across the load around a constant value v_d, which is equivalent to regulation of the capacitor voltage x_4 to the constant value $x_4^\star = v_d$. The admissible equilibria of the system (4.50) can be parameterized by the reference $x_4^\star$ as follows:

$$x^\star := \begin{bmatrix} \frac{1}{r_L (u^\star)^2} & \frac{1}{r_L u^\star} & u^\star & 1 \end{bmatrix}^\top x_4^\star,$$

where $u^\star = \sqrt{\frac{E}{x_4^\star}}$ is the corresponding constant control.

Applying the procedure of Proposition 4.7, we defined the output signal

$$y_N = -\sqrt{E v_d} x_1 - v_d x_2 + \frac{v_d^2}{E r_L} x_3 + \frac{v_d}{r_L} \sqrt{\frac{v_d}{E}} x_4. \tag{4.51}$$

It is easy to show that this output is zero-state detectable. That is, the following implication is true:

$$y_N(t) \equiv 0 \quad \Rightarrow \quad \lim_{t\to\infty} x(t) = x^\star.$$

Hence, the PI-PBC (4.47) ensures global asymptotic stability of the closed-loop.

Remark 4.16: It is important to underscore that the only parameters that are required for the implementation of the controller are r_L and E. Moreover, in Hernández-Gómez et al. (2010) a globally convergent adaptation scheme for r_L is proposed. The performance of the proposed controller is illustrated experimentally in Hernandez-Gomez et al. (2010).

Remark 4.17: The PI-PBC of Proposition 4.8 has been applied to a large variety of power converters in different scenarios. In Hernandez-Gomez et al. (2010) an adaptive PI-PBC for an AC–DC power converter with unknown resistance load is proposed. In Jaafar et al. (2013), a general framework to design immersion and invariance (Astolfi et al., 2008) observers for power converters controlled with the PI-PBC with partial state measurement is proposed. Experimental results for its application to the single-ended primary inductor converter are given. In Cisneros et al. (2015) and Bergna-Díaz et al. (2019), they are applied to modular multilevel converts and multiterminal HVDC systems, respectively.

4.5 PI-PBC of HVDC Power Systems

In this section, we propose to use the PI-PBC developed in Section 4.1 to design the inner-loop control of the HVDC system. Further details of the material reported in this section may be found in Zonetti et al. (2015) and Zonetti and Ortega (2015).

4.5.1 Background

We have witnessed in the last few years an ever widespread utilization of renewable energy utilities, mainly based on wind and solar power (Jager-Waldau, 2007; Chatzivasileiadis et al., 2013). Because of its intermittent nature the integration of this generating units to the existing AC distribution network poses a challenging problem (Carrasco et al., 2006; Lund, 2005). For this, and other reasons related to reduced losses and problems with reactive power and voltage stability in AC systems, the option of HVDC transmission systems is gaining wide popularity, see Chatzivasileiadis et al. (2013), Kirby et al. (2002), and Johansson et al. (2004) for additional motivations and details.

The main components of an HVDC system are AC to DC power converters, transmission lines, and voltage bus capacitors. The power converters connect the AC sources – that are associated to renewable generating units or to AC grids – to an HVDC grid through voltage bus capacitors. Two notable features distinguish HVDC systems from standard AC ones: the absence of loads and the central role played by the power converters, whose dynamics is highly *nonlinear*.

For its correct operation, HVDC systems – like all electrical power systems – must satisfy a large set of different regulation objectives that are, typically, associated with to the multiple time-scale behavior of the system. One way to deal with this issue, that prevails in practice, is the use of hierarchical architectures. These are nested control loops, at different time scales, each one providing references for an inner controller (Kazmierkowski et al., 2002; Yazdani and Iravani, 2010). In this section, we focus on the "innermost" control loop for HVDC transmission systems, that is, the control at the power converter level – in the sequel we will refer to this level as *inner-loop* control. The objective of the inner-loop control is to asymptotically drive the HVDC system toward a desired steady-state regime determined by the user. Regulation should be achieved selecting a suitable switching policy for the converters. A major practical constraint is that the control should be *decentralized*. That is, the controller of each power converter has only available for measurement its corresponding coordinates, with no exchange of information between them.

The main contributions reported in this chapter are as follows:

- **C1** Show that the HVDC system with a decentralized PI-PBC is GAS.
- **C2** Establish the close connections between the proposed PI and the popular Akagi's PQ instantaneous power method (Akagi, 2007). More specifically, we prove that driving the new passive output y_N to zero is tantamount to satisfying the power equalization objective of Akagi's method.
- **C3** We also show that using PI-PBC for this application has an important downside. Namely, that the zero dynamics in HVDC systems is "extremely slow," stymying the achievement of fast transient responses. On the other hand, it is also shown that other inner-loop PI controllers reported in the literature may exhibit unstable behavior because the zero dynamics associated with the corresponding outputs are *nonminimum phase*.

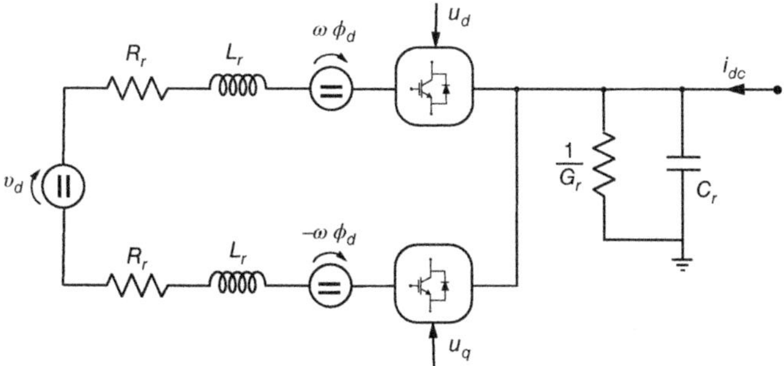

Figure 4.2 Schematic diagram of the equivalent circuit of a VSR in *dq* frame.

C4 To improve the transient performance, we follow common engineering practice and propose the addition of an outer-loop to define the PI-PBC reference signals – that is, the nested loop architecture known as *primary control.* The transient performance limitation of the PI-PBC, and its drastic improvement using primary control, is verified via simulations on a three-terminals benchmark example.

In this section, we limit our attention to the points **C1** and **C2** and refer the interested reader to Zonetti et al. (2015) and Zonetti and Ortega (2015) for the other two issues.

4.5.2 Port-Hamiltonian Model of the System

A pH model for the HVDC system, consisting of voltage source rectifiers (VSRs), transmission lines, and voltage bus capacitors, was reported in Zonetti et al. (2015), under the following standing assumptions, which are widely accepted in practice.

A1 Balanced operation of the three phase line voltages.
A2 Ideal four quadrant operation of the VSRs.
A3 Synchronized operation of the VSRs.[7]
A4 Meshed topology of the HVDC system where each VSR is connected to the ground and to a VSR-bus, while the lines directly connect VSR-buses, according to a determined meshed structure.

We consider in the section the well-known average model, expressed in dq-coordinates, of a single VSR shown in Figure 4.2, see Escobar et al. (1999), with the transmission lines represented by linear RL circuits.

---

7 Synchronized operation of the VSRs is usually achieved via robust phase-locked-loop detection of the latching frequencies (Yazdani and Iravani, 2010).

The pH model of the overall system with n VSRs and ℓ RL transmission lines takes the form

$$\dot{x} = [\mathcal{J}(u) - \mathcal{R}]\nabla H + EV, \tag{4.52}$$

with

- state $x(t) := \text{col}(x_R, x_L) \in \mathbb{R}^{3n+\ell}$, where $x_R(t)$ contains the inductors fluxes and the capacitors charges of the VSRs and $x_L(t)$ are the inductor fluxes of the lines;
- energy function $H(x) := \frac{1}{2}x^\top Qx$, with $Q = \text{diag}\{Q_R, Q_L\}$ a positive definite, diagonal matrix;
- duty cycles (controls) $u(t) = \text{col}(u_d(t), u_q(t)) \in \mathbb{R}^{2n}$;
- interconnection matrix.

$$\mathcal{J}(u) := \begin{bmatrix} J_R(u) & -E_3M \\ M^\top E_3^\top & 0 \end{bmatrix}, \tag{4.53}$$

where

$$J_R(u) := \sum_{i=1}^{n}(J_{R0,i}L_{r,i}\omega_i + J_{Rd,i}u_{d,i} + J_{Rq,i}u_{q,i}) \tag{4.54}$$

with ω_i the AC sides frequencies and $J_{R0,i}, J_{Rd,i}, J_{Rq,i}$ skew-symmetric matrices;
- dissipation matrix $\mathcal{R}$ diagonal and positive definite;
- input matrix $E := \begin{bmatrix} E_1^\top & 0 \end{bmatrix}^\top$.

The model is completed with the matrices $M \in \mathbb{R}^{n\times\ell}$, that is the incidence matrix of the subgraph obtained eliminating the VSRs edges and the ground node, $E_3 := \begin{bmatrix} 0 & 0 & I_n \end{bmatrix}^\top \in \mathbb{R}^{3n\times n}$ and $E_1 := \begin{bmatrix} I_n & 0 & 0 \end{bmatrix}^\top$. For further details on the model, the reader is referred to Zonetti et al. (2015).

4.5.3 Main Result

In Zonetti et al. (2015), it is shown that the assignable equilibria of the HVDC system (4.52) is obtained via the solution of n quadratic equations for the VSR coordinates. These equations are the well-known *power flow steady-state equations* of the individual VSR subsystems – from which we *univocally* determine the transmission lines coordinates equilibria. Once the desired equilibria is chosen, we invoke the construction of Proposition 4.7 to construct the shifted passive output and then, applying Proposition 4.8, design the GAS decentralized PI-PBC.

Proposition 4.9: *Consider the HVDC transmission system* (4.52). *Let $x^\star \in \mathcal{E}$ be the desired equilibrium with corresponding (univocally defined)* *control* $u^\star \in \mathbb{R}^{2n}$. *Define the output signal*

$$y_N := \begin{bmatrix} \text{col}\left(x_R^{\star\top} Q_R J_{Rd,i} Q_R x_R\right) \\ \text{col}\left(x_R^{\star\top} Q_R J_{Rq,i} Q_R x_R\right) \end{bmatrix} \in \mathbb{R}^{2n}. \tag{4.55}$$

The mapping $\tilde{u} \to y_N$ is passive. *More precisely, the system verifies the dissipation inequality*

$$\dot{H}_B \leq y^\top \tilde{u}, \tag{4.56}$$

with storage function $H_B(\tilde{x}) = \frac{1}{2}\tilde{x}^\top Q \tilde{x}$.

We are in position to present the main result of the section, whose proof follows from the previous material.

Proposition 4.10: *Consider the HVDC transmission system* (4.52), *with a desired steady-state* $x^\star \in \mathcal{E}$, *in closed-loop with the decentralized PI control* (4.47) *with* y_N *given in* (4.55) *and block diagonal, positive definite, gain matrices* K_P *and* K_I. *The equilibrium point* $(x^\star, K_I^{-1}u^\star)$ *is GAS.*

Remark 4.18: The proposed PI-PBC is decentralized in the sense that, for its implementation, each VSR control requires only the measurement of its corresponding inductor currents and capacitor voltage. Guaranteeing this property motivates our choice of block diagonal gain matrices in the PI-PBC.

4.5.4 Relation of PI-PBC with Akagi's PQ Method

A dominant approach for the design of controllers for reactive power compensation using active filters (for three-phase circuits) is the PQ instantaneous power method proposed by Akagi et al. in Akagi (2007). It consists in an outer-loop that generates references for the inner PI loops. The references are selected in order to satisfy a very simple heuristic: the AC active power P has to be equal to the DC power P_{dc}, thus ensuring the maximal power transfer from AC to DC side, and the reactive power should take a desired value.

To compare the PI-PBC with Akagi's controller, let us write the passive output (4.55) in coenergy variables for a single VSR as

$$y_N = \begin{bmatrix} v_C^\star i_d - i_d^\star v_C \\ v_C^\star i_q - i_q^\star v_C \end{bmatrix}, \tag{4.57}$$

where we recall that $(i_d^\star, i_q^\star, v_C^\star) \in \mathcal{E}$, that is, they belong to the assignable equilibrium set. Let us, then, define the active AC and DC powers at the equilibrium as

$$P^\star = v_d i_d^\star, \quad P_{dc}^\star = v_C^\star i_{dc}.$$

Consider then the following equivalences:

$$P^{\star}P_{dc} = P^{\star}_{dc}P \Leftrightarrow v^{\star}_{C}i_d = i^{\star}_{d}v_C \Leftrightarrow e_1^{\top}y_N = 0. \tag{4.58}$$

Similarly, for the reactive power,

$$Q^{\star}P_{dc} = P^{\star}_{dc}Q \Leftrightarrow v^{\star}_{C}i_q = i^{\star}_{q}v_C \Leftrightarrow e_2^{\top}y_N = 0. \tag{4.59}$$

In other words, the objective of the PI-PBC to drive the passive output y_N to zero can be reinterpreted as a power equalization objective identical to the one used in Akagi's PQ method.

4.6 PI-PBC of Wind Energy Systems

In this section, we design a PI-PBC for a wind energy production system with guaranteed stability properties. Providing a theoretical framework for the controller design allows power engineer practitioners to apply the control law with confidence and considerably simplifies the commissioning stage – hence, is a topic of paramount importance. Further details of the material reported in this section may be found in Cisneros et al. (2016) and Cisneros et al. (2020).

4.6.1 Background

Wind systems are composed of a mechanical subsystem, a turbine, and an electrical subsystem that consists of a generator and a power converter. The overall dynamics is complicated, highly nonlinear and uncertain and with limited control authority, making most model-based nonlinear controller reported in the literature (Isidori, 1995; Krstić et al., 1995; Sepulchre et al., 2011) practically unfeasible.

To simplify the controller design, it is common to dissociate the generator/converter control and the turbine dynamics control. This dissociation is rationalized invoking the difference of their time responses – being the electrical part faster than the mechanical one. Regarding turbine control, the methods most frequently used in industry rely on classical linear systems analysis, which are based on the linearization of the nonlinear dynamics, see Leithead and Connor (2013) for a review of these techniques. Classical backstepping methods have been used in Galeazzi et al. (2015, 2013), and Lu and Lin (2011), while Calderaro et al. (2008) and Bianchi et al. (2007) address the problem using fuzzy and gain-scheduling controllers, respectively. Standard nonlinear control methods (Isidori, 1995), based on feedback linearization, have been used in Boukhezzar and Siguerdidjane

(2011). As mentioned above, all these works focus on the turbine control, making the assumption that the faster electrical part is ideally controlled by means of an inner loop whose references are provided by an outer loop involving the mechanical part. Therefore, there is no guarantee that the overall system will perform well or even preserve stability.

Controllers considering the whole system have also been reported. See, for example, Boukezzar and M'Saad (2008) for a sliding mode control of a wind system based on a fully actuated – hence, easier to control – doubly fed induction generator. In Valenciaga et al. (2003), control of a solar and wind generation system to satisfy a power demand is accomplished using sliding mode control. By cancelling undesirable nonlinearities, this high-gain control renders the system passive respect to an energy function which depends on the sliding surface. In Salazar et al. (2012) a control of a variable speed wind turbine emulator with a permanent magnet synchronous generator (PMSG) employing standard PI controllers with partial nonlinearity decoupling is presented. As shown in Cisneros et al. (2013), these PI controllers, besides having no stability guarantee, are difficult to tune and exhibit poor robustness properties.

In this section, we propose to use the PI-PBC developed in Proposition 4.8. Toward this end, a passive output for the nonlinear incremental model is identified around which the stabilizing PI is added. Interestingly, as shown in Section 4.5, the PI scheme that results from the application of this method is closely related to the well-known instantaneous active power control of Akagi (2007). In this way, an important connection with current practice is established. Particularly, one of the advantages of the proposed controller is related to its simplicity as the resulting controller is a linear PI.

4.6.2 System Model

Depicted in Figure 4.3, the system consists of a turbine, surface-mounted PMSG, and a rectifier connected to a load, a capacitor, a load, and one constant voltage source, which is used to form the DC bus.

Wind Turbine

The mechanical power extracted from the wind is given by the power function:

$$P_w = \frac{1}{2}\rho A C_p(\lambda) v_w^3, \tag{4.60}$$

where v_w is the wind speed, which is assumed constant and known, ρ is the air density, A is the area swept by the blades, and C_p is the turbine's

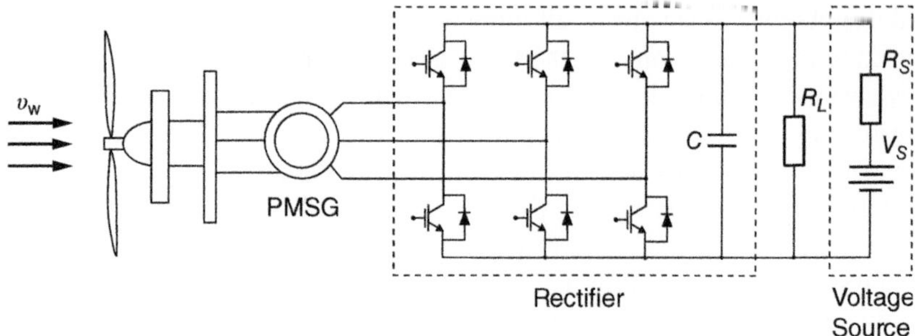

Figure 4.3 Wind energy system.

coefficient power, which is function of the tip-speed ratio λ defined as

$$\lambda := \frac{r\omega_m}{v_w}, \tag{4.61}$$

where ω_m is the shaft's rotational speed, and r is the blade's radius.

The turbine's power coefficient also depends on the blade pitch angle. However, in this section, it is assumed that the turbine operates in the regime where this angle is kept constant, also known as Region 2 (Pao and Johnson, 2009). For this reason, this variable has been omitted in the definition.

Maximum Power Extraction

A typical shape of the function C_p is depicted in Figure 4.4. It can be seen from the figure that we define

$$\lambda^\star = \arg\max C_p(\lambda). \tag{4.62}$$

The turbine is required to operate at this value to maximize the power extracted from the wind. Using (4.61), it is possible to obtain the corresponding optimal value for the schaft's rotational speed, i.e.

$$\omega_m^\star := \frac{v_w \lambda^\star}{r}. \tag{4.63}$$

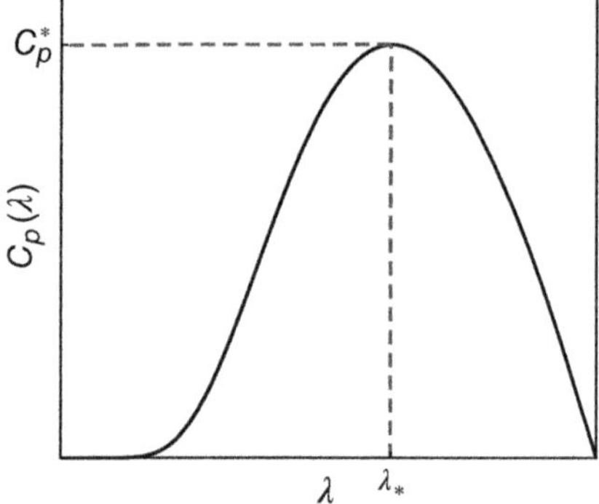

Figure 4.4 Function $C_p(\lambda)$.

Turbine Model

The dynamic equation of a one-mass turbine is obtained from Newton's equation of moton:

$$J\dot{\omega}_m = -f\omega_m + T_m - T_e, \tag{4.64}$$

where J is the rotor inertia, $f > 0$ is a friction coefficient, T_m is the mechanical torque applied to the windmill shaft

$$\begin{aligned} T_m = \frac{P_w}{\omega_m} &= \frac{1}{2}\rho A r v_w^2 \frac{C_p(\lambda)}{\lambda} \\ &= \frac{1}{2}\rho A v_m^3 \frac{1}{\omega_m} C_p\left(\frac{r\omega_m}{v_w}\right), \end{aligned} \tag{4.65}$$

and T_e, to be defined, is the electrical torque provided by the generator.

PMSG and Electrical Part

The dynamic equations of the PMSG in dq-coordinates are

$$\begin{aligned} L\dot{i}_d &= -Ri_d + Li_q\omega_e - v_d, \\ L\dot{i}_q &= -Ri_q - Li_d\omega_e + \phi\omega_e - v_q, \end{aligned} \tag{4.66}$$

where $i_q(t), i_d(t), v_q(t), v_d(t)$ are, respectively, the q and d components of the current and voltage, R and L are the stator resistance and inductance, respectively, ϕ is the permanent magnetic flux and ω_e is the electrical frequency. The electrical frequency satisfies the relation

$$\omega_e = \frac{P}{2}\omega_m, \tag{4.67}$$

where P is the number of pole pairs. The electrical torque T_e is given by

$$T_e = \frac{3}{2}\frac{P}{2}\phi i_q. \tag{4.68}$$

The input voltages in the generator are

$$v_d = v_C d_1, \quad v_q = v_C d_2, \tag{4.69}$$

where d_1 and d_2 are duty ratio of the rectifier control signals in dq-coordinates. Finally, from Kirchhoff's current law,

$$C\dot{v}_C = -R_e v_C + \frac{V_S}{R_S} + i_d d_1 + i_q d_2, \tag{4.70}$$

where C is the capacitance value, $R_e := \frac{R_L+R_S}{R_L R_S}$, R_L is a resistive load, and R_S and V_S are, respectively, the supply internal resistance and dc voltage.

The Overall System

Substituting (4.65) and (4.68) in (4.64), (4.69) and (4.67) in (4.66) and from (4.70), the overall system becomes

$$L\dot{i}_d = -Ri_d + \frac{P}{2}Li_q\omega_m - d_1 v_C, \tag{4.71}$$

$$L\dot{i}_q = -Ri_q - \frac{P}{2}Li_d\omega_m + \frac{P}{2}\phi\omega_m - d_2 v_C, \tag{4.72}$$

$$J\dot{\omega}_m = -f\omega_m + \frac{1}{2}\rho A v_m^3 \frac{1}{\omega_m} C_p\left(\frac{v_w\omega_m}{r}\right) - \frac{3}{2}\frac{P}{2}\phi i_q, \tag{4.73}$$

$$C\dot{v}_C = -R_e v_C + \frac{V_S}{R_S} + d_1 i_d + d_2 i_q. \tag{4.74}$$

4.6.3 Control Problem Formulation

The control objectives are

O1 Minimize the copper loss and maximize the efficiency of the generator, i.e. to make the system operate at the i_d current value $i_d = i_{d\star} = 0$.
O2 Operate at the maximum power extraction point $\omega_m = \omega_m^\star = \frac{v_w\lambda^\star}{r}$.

In the next lemma, the equilibrium points, compatible with the aforementioned control objectives, are stated.

Lemma 4.2: *The assignable equilibrium points of the system* (4.71)–(4.74) *are defined by the set*

$$\mathcal{E} := \left\{ (i_d, i_q, \omega_m, v_C) | i_d = 0, \omega_m = \omega_m^\star, i_q = \frac{4}{3P\phi}(T_m^\star - f\omega_m^\star), \right.$$
$$\left. h(v_C) = 0 \right\}, \tag{4.75}$$

where $T_m^\star := T_m(\omega_m^\star)$ *and*

$$h_1(v_C) := R_e v_C^2 - \frac{V_S}{R_S}v_C + \frac{16R}{9P^2\phi^2}\left(T_m^\star - f\omega_m^\star\right)^2 + \frac{2}{3}f\omega_m^{\star 2} - \frac{2}{3}\omega_m^\star T_m^\star.$$

Proof. Consider that $z = \mathrm{col}(i_d, i_q, \omega_m, v_C)$. Then, at the equilibrium point, the system (4.71)–(4.74) can be written as

$$0 = \varphi(z) + g(z)d, \tag{4.76}$$

where

$$\varphi(z) := \begin{bmatrix} -Ri_d + \frac{P}{2}Li_q\omega_m \\ -Ri_q - \frac{P}{2}Li_d\omega_m + \frac{P}{2}\phi\omega_m \\ -\frac{2}{3}f\omega_m + \frac{2}{3}T_m - \frac{P}{2}\phi i_q \\ -R_e v_C + \frac{V_S}{R_S} \end{bmatrix}, \quad g(z) := \begin{bmatrix} -v_C & 0 \\ 0 & -v_C \\ 0 & 0 \\ i_d & i_q \end{bmatrix},$$

$$d := \begin{bmatrix} d_1 \\ d_2 \end{bmatrix}. \tag{4.77}$$

The full-rank left-annihilator of $g(z)$ is the matrix

$$g^{\perp}(z) = \begin{bmatrix} z^{\top} \\ e_3^{\top} \end{bmatrix}, \tag{4.78}$$

where $e_3 = \text{col}(0, 0, 1, 0)$. Then, multiplying (4.76) by $g^{\perp}(z)$ yields

$$0 = g^{\perp}(z)\varphi(z). \tag{4.79}$$

Fixing $i_d = 0$ and $\omega_m = \omega_m^{\star}$, the last equation is equivalent to

$$0 = \begin{bmatrix} -Ri_q^2 - \frac{2}{3}f\omega_m^{\star 2} + \frac{2}{3}T_m^{\star}\omega_m^{\star} - R_e v_C^2 + \frac{V_S}{R_S}v_C \\ -\frac{2}{3}f\omega_m^{\star} + \frac{2}{3}T_m^{\star} - \frac{P}{2}\phi i_q \end{bmatrix}. \tag{4.80}$$

From the second equation, it follows that

$$i_q = \frac{4}{3P\phi}(T_m^{\star} - f\omega_m^{\star}).$$

Finally, substituting the latter equation in the first equation of (4.80) yields the last algebraic equation of the assignable equilibrium set. □

Remark 4.19: The necessary and sufficient condition for the existence of the desired equilibrium point is

$$\frac{V_S^2}{4R_e R_S^2} \geq \frac{16R}{9P^2\phi^2}(T_m^{\star} - f\omega_m^{\star})^2 + \frac{2}{3}f\omega_m^{\star} - \frac{2}{3}\omega_m^{\star}T_m^{\star}. \tag{4.81}$$

4.6.4 Proposed PI-PBC

To proceed with the design of the PI-PBC, the system is written in the pH representation. For, the following change of variables is considered

$$x = \text{col}(Li_d, Li_q, J_1\omega_m, Cv_C), \quad u = \text{col}(d_1, d_2). \tag{4.82}$$

Then, the system becomes

$$\begin{aligned} \dot{x}_1 &= -\frac{R}{L}x_1 + \frac{\eta}{J_1}x_2x_3 - \frac{x_4}{C}u_1, \\ \dot{x}_2 &= -\frac{\eta}{J_1}x_1x_3 - \frac{R}{L}x_2 + \frac{\phi_1}{J_1}x_3 - \frac{x_4}{C}u_2, \\ \dot{x}_3 &= -f_1x_3 + \Phi(x_3) - \frac{\phi_1}{L}x_2, \\ \dot{x}_4 &= -\frac{R_e}{C}x_4 + \frac{x_1}{L}u_1 + \frac{x_2}{L}u_2 + \frac{V_S}{R_S}, \end{aligned} \tag{4.83}$$

where

$$\Phi(x_3) := \frac{1}{3}\rho AJ_1v_w^3\frac{1}{x_3}C_p\left(\frac{rx_3}{J_1v_w}\right)$$

and

$$\phi_1 := \frac{\phi P}{2}, \quad \eta := \frac{P}{2}, \quad J_1 := \frac{2}{3}J, \quad f_1 := \frac{2}{3}f. \tag{4.84}$$

The pH representation of the system is

$$\dot{x} = \left[J_0(x_3) - \mathcal{R} + J_1 u_1 + J_2 u_2\right] \nabla H(x) + E_0(x_3), \tag{4.85}$$

where $H(x) := \frac{1}{2}x^\top Q x$ is the total energy of the system, the external force E_0 is defined as

$$E_0(x_3) := \mathrm{col}\left(0, 0, \Phi(x_3), \frac{V_S}{R_S}\right)$$

and the matrices $Q := \mathrm{diag}\left(\frac{1}{L}, \frac{1}{L}, \frac{1}{J_1}, \frac{1}{C}\right)$, $\mathcal{R} := \mathrm{diag}\left(R, \; R, f_1, R_e\right)$,

$$J_0(x_3) := \begin{bmatrix} 0 & \frac{\eta}{J_1}Lx_3 & 0 & 0 \\ -\frac{\eta}{J_1}Lx_3 & 0 & \phi_1 & 0 \\ 0 & -\phi_1 & 0 & 0 \\ 0 & 0 & 0 & 0 \end{bmatrix}, \quad J_1 := \begin{bmatrix} 0 & 0 & 0 & -1 \\ 0 & 0 & 0 & 0 \\ 0 & 0 & 0 & 0 \\ 1 & 0 & 0 & 0 \end{bmatrix},$$

$$J_2 := \begin{bmatrix} 0 & 0 & 0 & 0 \\ 0 & 0 & 0 & -1 \\ 0 & 0 & 0 & 0 \\ 0 & 1 & 0 & 0 \end{bmatrix}.$$

To carry out the analysis, we make the usual assumption that the mechanical dynamics is much slower than the electrical one and suppose that the generator speed ω_m is well-regulated. This can be translated into the following standing assumption.

Assumption 4.4: The system dynamics is represented by

$$\dot{x} = (J - \mathcal{R} + J_1 u_1 + J_2 u_2)Qx + E, \tag{4.86}$$

where $J := J_0(x_3^\star)$ and $E := E_0(x_3^\star)$.

Notice that x_3 is assumed to be constant only when it appears in the interconnection matrix J_0 and the external force E_0, but it remains a state variable in the overall dynamics.

Applying the derivations of Section 4.4 and doing some standard analysis of cascaded systems we obtain the following result, whose proof may be found in Cisneros et al. (2016), where some realistic simulations are also carried out.

Proposition 4.11: *Consider the system (4.86) in closed-loop with the PI controller (4.47) where y_N is given by*

$$y_N = \begin{bmatrix} i_d^\star \tilde{v}_C - \tilde{i}_d v_C^\star \\ i_q^\star \tilde{v}_C - \tilde{i}_q v_C^\star \end{bmatrix}. \tag{4.87}$$

For all initial conditions $(x(0), x_c(0))$, the trajectories of the closed-loop system are bounded and $\lim_{t\to\infty} \tilde{x}(t) = 0$.

Remark 4.20: In Cisneros et al. (2020), we have been able to relax the time-scale separation Assumption 4.4 via the addition of a dynamics standard PBC similar to the ones reported in Ortega et al. (1998).

Remark 4.21: The research reported in this section was carried out within the framework of the FREEDM System Center (http://www.freedm.ncsu.edu/) whose main objective is the implementation and testing of a solid-state transformer enabled microgrid. The grid possesses distributed renewable energy resources such as the wind generation unit considered here. Current work is under way for the experimental implementation of the proposed PI-PBC.

4.7 Shifted Passivity of PI-Controlled Permanent Magnet Synchronous Motors

In this section, we prove that the classical approach to control the current of PMSMs – via a simple PI – yields a GAS closed-loop. Toward this end, we invoke Proposition 4.6 to prove a key shifted passivity property of the PMSM. Further details of the material reported in this section may be found in Ortega et al. (2018).

4.7.1 Background

Control of electric motors is achieved in the vast majority of commercial drives via nested loop PI controllers (Leonhard, 1985; Krause et al., 2013): the inner one wrapped around current errors and an external one that defines the desired values for these currents to generate a desired torque – for speed or position control. The rationale to justify this control configuration relies on the, often reasonable, assumption of time-scale separation between the electrical and the mechanical dynamics. In spite of its enormous success, to the best of our knowledge, a rigorous theoretical analysis of the stability of this scheme has not been reported. The main contribution of this section is to (partially) fill-up this gap for the widely popular PMSM, proving that the inner-loop PI controller ensures global asymptotic stability of the closed-loop, provided some viscous friction (possibly arbitrarily small) is present in the rotor dynamics, that the load torque is known and the proportional gain of the PI is suitably chosen, i.e. sufficiently high.

Several globally stable position and velocity controllers for PMSMs have been reported in the control literature – even in the sensorless context, e.g. Bodson et al. (1993), Lee et al. (2010), Tomei and Verrelli (2008), Tomei and Verrelli (2011), and references therein. However, these controllers have received an, at best, lukewarm reception within the electric drives community, which overwhelmingly prefers the aforementioned nested-loop PI configuration. Several versions of PI schemes based on fuzzy control, sliding modes or neural network control have been intensively studied in applications journals, see Jung et al. (2015) for a recent review of this literature. To the best of our knowledge, a rigorous stability analysis of all these schemes is conspicuous by its absence. Various attempts to establish the result reported here have been published in the literature either relying on linear approximations of the motor dynamics or including additional terms that cancel some nonlinear terms, see Hernández-Guzmán and Carrillo-Serrano (2011) and Hernández-Guzmán and Silva (2011) and references therein – a standing assumption being, similarly to us, the existence of viscous friction.

4.7.2 Motor Models

In this section we present the motor model, define the desired equilibrium and give its incremental model.

Standard dq Model

The dynamics of the surface-mounted PMSM in the dq frame is described by Krause et al. (2013):

$$\begin{aligned} L_d \frac{di_d}{dt} &= -R_s i_d + \omega L_q i_q + v_d, \\ L_q \frac{di_q}{dt} &= -R_s i_q - \omega L_d i_d - \omega\Phi + v_q, \\ J\frac{d\omega}{dt} &= -R_m\omega + n_p\left[(L_d - L_q) i_d i_q + \Phi i_q\right] - \tau_L, \end{aligned} \tag{4.88}$$

where $i_d(t), i_q(t)$ are currents, $v_d(t), v_q(t)$ are voltage inputs, ω is the electrical angular velocity,[8] $\frac{2n_p}{3}$ is the number of pole pairs, $L_d > 0, L_q > 0$ are the stator

8 Related with the rotor speed ω_m via $\omega = \frac{2n_p}{3}\omega_m$.

inductances, $\Phi > 0$ is the back emf constant, $R_s > 0$ is the stator resistance, $R_m > 0$ is the viscous friction coefficient, $J > 0$ is the moment of inertia and τ_L is a constant load torque.

Defining the state and control vectors as

$$x := \begin{bmatrix} i_d \\ i_q \\ \omega \end{bmatrix}, \quad u := \begin{bmatrix} v_d \\ v_q \end{bmatrix}$$

the system (4.88) can be written in compact form as

$$\mathcal{D}\dot{x} + [C(x) + \mathcal{R}]x = Gu + d,$$

where

$$\mathcal{D} := \begin{bmatrix} L_d & 0 & 0 \\ 0 & L_q & 0 \\ 0 & 0 & \frac{J}{n_p} \end{bmatrix} > 0, \quad \mathcal{R} := \begin{bmatrix} R_s & 0 & 0 \\ 0 & R_s & 0 \\ 0 & 0 & \frac{R_m}{n_p} \end{bmatrix} > 0,$$

$$C(x) := \begin{bmatrix} 0 & 0 & -L_q x_2 \\ 0 & 0 & L_d x_1 + \Phi \\ L_q x_2 & -(L_d x_1 + \Phi) & 0 \end{bmatrix} = -C^\top(x),$$

$$G := \begin{bmatrix} 1 & 0 \\ 0 & 1 \\ 0 & 0 \end{bmatrix}, \quad d := \begin{bmatrix} 0 \\ 0 \\ -\frac{\tau_L}{n_p} \end{bmatrix}.$$

Besides simplifying the notation, the interest of the representation above is that it reveals the power balance equation of the system. Indeed, the total energy of the motor is $H(x) = \frac{1}{2}x^\top \mathcal{D}x$, whose derivative yields

$$\underbrace{\dot{H}}_{\text{stored power}} = -\underbrace{x^\top \mathcal{R}x}_{\text{dissipated}} + \underbrace{y^\top u}_{\text{supplied}} - \underbrace{x_3 \frac{\tau_L}{n_p}}_{\text{extracted}}, \tag{4.89}$$

where we used the skew-symmetry of $C(x)$ and defined the currents as outputs, that is,

$$y := G^\top x = \begin{bmatrix} i_d \\ i_q \end{bmatrix}.$$

Incremental Model

The industry standard desired equilibrium is the maximum torque per ampere value defined as

$$x^\star := \text{col}\left(0, \frac{1}{n_p\Phi}(\tau_L + R_m\omega^\star), \omega^\star\right), \tag{4.90}$$

where $\omega^\star$ is the desired electrical speed. With respect to this equilibrium, we define the incremental model

$$\begin{aligned} \mathcal{D}\dot{\tilde{x}} + C(x)\tilde{x} + [C(x) - C^\star]x^\star + \mathcal{R}\tilde{x} &= G\tilde{u}, \\ \tilde{y} &= G^\top \tilde{x}, \end{aligned} \tag{4.91}$$

where $\widetilde{(\cdot)} := (\cdot) - (\cdot)^\star$, $C^\star := C(x^\star)$, and we used the fact that

$$\begin{aligned} (C^\star + \mathcal{R})x^\star &= Gu^\star + d, \\ y^\star &= G^\top x^\star, \end{aligned}$$

with

$$u^\star = \begin{bmatrix} -\frac{1}{n_p\Phi} L_q \omega^\star (\tau_L + R_m \omega^\star) \\ \Phi\omega^\star + \frac{1}{n_p\Phi} R_s (\tau_L + R_m \omega^\star) \end{bmatrix}.$$

Note that

$$\tilde{y} = \begin{bmatrix} x_1 \\ x_2 - \frac{1}{n_p\Phi}(\tau_L + R_m\omega^\star) \end{bmatrix}. \tag{4.92}$$

4.7.3 Problem Formulation

We are interested in the section in giving conditions for global asymptotic stability of a PI controller wrapped around the currents $i_d(t), i_q(t)$, which are assumed to be measurable. We assume known τ_L, Φ and R_m and design a "classical" PI

$$\begin{aligned} \dot{x}_c &= \tilde{y}, \\ u &= -K_I x_c - K_P \tilde{y}, \end{aligned} \tag{4.93}$$

with $\tilde{y}$ defined in (4.92) and $K_I, K_P > 0$.

We want to prove that there exists a positive-definite gain matrix $K_P^{\mathtt{min}}$ such that the PMSM model (4.88) in closed-loop with the PI controller (4.93) with $K_P \geq K_P^{\mathtt{min}}$ has a GAS equilibrium.

Remark 4.22: The assumption of known load torque is relaxed in Ortega et al. (2018) proposing an adaptive scheme that, in the spirit of the aforementioned outer-loop PI, generates, via the addition of a simple integrator, an estimate for it – preserving global asymptotic stability of the new scheme.

Remark 4.23: As indicated in the introduction, in practice the reference value for x_2 is generated with an outer-loop PI around speed errors, that is,

$$\begin{aligned} \dot{\chi} &= \tilde{x}_3, \\ \hat{x}_2^\star &= -a_I \chi - a_P \tilde{x}_3, \end{aligned} \tag{4.94}$$

with $a_I, a_P > 0$. Unfortunately, the stability analysis of this configuration is far from obvious.

4.7.4 Main Result

In this section, we give our main result, that is conditions under which, the PMSM model (4.88) in closed-loop with the PI controller (4.93) yields a GAS closed-loop. The key step is to prove, using Proposition 4.6, that the PMSM is shifted passifiable, a result contained in Lemma 4.3.

Lemma 4.3: *Define the matrix*

$$\mathcal{B} := \begin{bmatrix} 2R_s + 2\epsilon & (L_d - L_q)x_3^\star & -L_d x_2^\star \\ (L_d - L_q)x_3^\star & 2R_s + 2\epsilon & 0 \\ -L_d x_2^\star & 0 & 2\frac{R_m}{n_p} \end{bmatrix},$$

for some $\epsilon \in \mathbb{R}$. *If* $\mathcal{B} \geq 0$ *the PMSM satisfies*

$$\dot{H}_B \leq \epsilon |\tilde{y}|^2 + \tilde{y}^\top \tilde{u} \tag{4.95}$$

with $H_B(\tilde{x}) = \frac{1}{2}\|\tilde{x}\|_D^2$.

We are in a position to present our main result, whose proof follows immediately from previous material – see also Ortega et al. (2018).

Proposition 4.12: *Consider the PMSM model* (4.88) *in closed-loop with the PI controller* (4.93), *the integral gain* $K_I > 0$, *and the proportional gain*[9] $K_P = k_p I_2 > 0$. *There exists a positive constant* k_p^{min} *such that*

$$k_p \geq k_p^{min} \tag{4.96}$$

ensures that $(x^\star, x_c^\star)$ *is a GAS equilibrium of the closed-loop system. For non-salient PMSM, i.e. when* $L_d = L_q$, *the constant* k_p^{min} *can be chosen such that*

$$k_p^{min} > \frac{L_d^2}{4R_m n_p \Phi^2}\left(\tau_L + R_m|\omega^\star|\right)^2 - R_s. \tag{4.97}$$

4.7.5 Conclusions and Future Research

We have established the practically interesting – though not surprising – result that the PMSM can be globally regulated around a desired equilibrium

9 K_p is taken of this particular form to simplify the presentation of the main result – this choice is done without loss of generality.

point with a simple PI control around the current errors, provided some viscous friction is present in the rotor dynamics and the proportional gain of the PI is suitably chosen. The key ingredient to establish this result is the property of shifted passifiability stated in Lemma 4.1.

From the theoretical viewpoint, the main drawback of the results reported in the section are the requirement of existence, and knowledge, of the friction coefficient R_m. As shown in Ortega et al. (2018), the requirement of knowing R_m can be relaxed – at the price of complicating the controller and requiring some excitation conditions. However, the assumption of $R_m > 0$ seems unavoidable if we want to preserve a simple PI structure. It should be underscored, however, that from the practical viewpoint, the assumption that the mechanical dynamics has some static friction – that may be arbitrarily small – is far from being unreasonable.

Bibliography

A. Akagi. *Instantaneous Power Theory and Applications to Power Conditioning*. Wiley, 2007.

A. Alonso and B. E. Ydstie. Stabilization of distributed systems using irreversible thermodynamics. *Automatica*, 37(11): 1739–1755, 2001.

S. Aranovskiy, R. Ortega, and R. Cisneros. A robust PI passivity-based control of nonlinear systems and its application to temperature regulation. *International Journal of Robust and Nonlinear Control*, 26(10): 2216–2231, 2016.

A. Astolfi, D. Karagiannis, and R. Ortega. *Nonlinear and Adaptive Control with Applications*. Springer-Verlag, London, 2008.

G. Bergna-Díaz, D. Zonetti, S. Sánchez, R. Ortega, and E. Tedeschi. PI passivity-based control and performance analysis of MMC multi-terminal HVDC systems. *IEEE Journal of Emerging and Selected Topics in Power Electronics*, 7(6): 2453–2477, 2019.

F. D. Bianchi, H. De Bautista, and R. J. Mantz. *Wind Turbine Control Systems: Principles, Modeling and Gain Scheduling Design*. Springer, 2007.

M. Bodson, J. N. Chiasson, R. T. Novotnak, and R. B. Rekowski. High-performance nonlinear feedback control of a permanent magnet stepper motor. *Transactions on Control Systems Technology*, 1(1): 5–14, 1993.

B. Boukezzar and M. M'Saad. Robust sliding mode control of a DFIG variable speed wind turbine for power production optimization. In *Mediterranean Conference on Control and Automation*, pages 795–800, 2008.

B. Boukhezzar and H. Siguerdidjane. Nonlinear control of a variable-speed wind turbine using a two-mass model. *IEEE Transactions on Energy Conversion*, 26(1): 149–162, 2011.

L. M. Bregman. The relaxation method of finding the common point of convex sets and its application to the solution of problems in convex programming. *USSR Computational Mathematics and Mathematical Physics*, 7(3): 200–217, 1967.

M. Bürger, D. Zelazo, and F. Allgöwer. Duality and network theory in passivity-based cooperative control. *Automatica*, 50(8): 2051–2061, 2014.

V. Calderaro, V. Galdi, A. Piccolo, and P. Siano. A fuzzy controller for maximum energy extraction from variable speed wind power generation systems. *Electric Power Systems Research*, 78(6): 1109–1118, 2008.

J. M. Carrasco, L. G. Franquelo, J. T. Bialasiewicz, E. Galván, R. C. P. Guisado, M. A. M. Prats, J. I. León, and N. Moreno-Alfonso. Power-electronic systems for the grid integration of renewable energy sources: a survey. *IEEE Transactions on Industrial Electronics*, 53(4): 1002–1016, 2006.

F. Castaños, B. Jayawardhana, R. Ortega, and E. García-Canseco. Proportional plus integral control for set point regulation of a class of nonlinear RLC circuits. *Circuits, Systems and Signal Processing*, 28(4): 609–623, 2009.

S. Chatzivasileiadis, D. Ernst, and G. Andersson. The global grid. *Renewable Energy*, 57: 372–383, 2013.

R. Cisneros, R. Gao, R. Ortega, and I. Husain. PI passivity-based control for maximum power extraction of a wind energy system with guaranteed stability properties. *International Journal of Emerging Electric Power Systems*, 17(5): 567–573, 2016.

R. Cisneros, R. Gao, R. Ortega, and I. Husain. A PI+passivity-based control of a wind energy conversion system enabled with a solid state transformer. *International Journal of Control*, 2020.

R. Cisneros, F. Mancilla-David, and R. Ortega. Passivity-based control of a grid-connected small-scale windmill with limited control authority. *IEEE Journal of Emerging and Selected Topics in Power Electronics*, 1(4): 247–259, 2013.

R. Cisneros, M. Pirro, G. Bergna-Díaz, R. Ortega, G. Ippoliti, and M. Molinas. Global tracking passivity-based PI control of bilinear systems and its application to the boost and modular multilevel converters. *Control Engineering Practice*, 43(10): 109–119, 2015.

C. de Persis, T. Noergard, R. Ortega, and R. Wisniewski. Output regulation of large scale hydraulic networks. *IEEE Transactions on Control Systems Technology*, 22(1): 238–245, 2014.

C. A. Desoer and M. Vidyasagar. *Feedback Systems: Input-Output Properties*. Academic Press, New York, 2009.

G. Escobar, A. J. van der Schaft, and R. Ortega. A Hamiltonian viewpoint in the modeling of switching power converters. *Automatica*, 35(3): 445–452, 1999.

J. Ferguson, R. H. Middleton, and A. Donaire. Disturbance rejection via control by interconnection of port-Hamiltonian systems. In *IEEE Conference on Decision and Control (CDC)*, pages 507–512, Osaka, Japan, 2015.

R. Galeazzi, K. T. Borup, H. Niemann, N. K. Poulsen, and F. Caponetti. Adaptive backstepping control of lightweight tower wind turbine. In *American Control Conference*, pages 3058–3065, Paris, France, 2015.

R. Galeazzi, M. P. S. Gryning, and M. Blanke. Observer backstepping control for variable speed wind turbine. In *American Control Conference*, pages 1036–1043, Washington, DC, USA, 2013.

V. M. Hernández-Guzmán and R. V. Carrillo-Serrano. Global PID position control of PM stepper motors and PM synchronous motors. *International Journal of Control*, 84(11): 1807–1816, 2011.

V. M. Hernández-Guzmán and R. Silva. PI control plus electric current loops for PM synchronous motors. *IEEE Transactions on Control Systems Technology*, 19(4): 868–873, 2011.

M. Hernández-Gómez, R. Ortega, F. Lamnabhi-Lagarrigue, and G. Escobar. Adaptive PI stabilization of switched power converters. *IEEE Transactions on Control Systems Technology*, 18(3): 688–698, 2010.

M. Hernández-Gómez, R. Ortega, F. Lamnabhi-Lagarrigue, and G. Escobar. *Dynamics and Control of Switched Electronic Systems*, pages 355–390. Advances in Industrial Control. Springer, Berlin/Heidelberg, 2012.

D. J. Hill and P. J. Moylan. Dissipative dynamical systems: basic input-output and state properties. *Journal of the Franklin Institute*, 309(5): 327–357, 1980.

G. H. Hines, M. Arcak, and A. K. Packard. Equilibrium-independent passivity: a new definition and numerical certification. *Automatica*, 47(9): 1949–1956, 2011.

A. Isidori. *Nonlinear Control Systems*. Springer, 1995.

A. Jaafar, A. Allawieh, R. Ortega, and E. Godoy. PI stabilization of power converters with partial state measurements. *IEEE Transactions on Control Systems Technology*, 21(2): 560–568, 2013.

A. Jager-Waldau. Photovoltaics and renewable energies in Europe. *Renewable and Sustainable Energy Reviews*, 11(7): 1414–1437, 2007.

B. Jayawardhana, R. Ortega, E. García-Canseco, and F. Castaños. Passivity of nonlinear incremental systems: Application to PI stabilization of nonlinear RLC circuits. *System & Control Letters*, 56(9–10): 618–622, 2007.

S. G. Johansson, G. Asplund, E. Jansson, and R. Rudervall. Power system stability benefits with VSC DC-transmission systems. In *CIGRE Conference*, Paris, France, 2004.

J. Jung, V. Leu, T. Do, E. Kim, and H. Choi. Adaptive PID speed control design for permanent magnet synchronous motor drives. *IEEE Transactions on Power Electronics*, 30(2): 900–908, 2015.

M. P. Kazmierkowski, R. Krishnan, F. Blaabjerg, and J. D. Irwin. *Control in Power Electronics: Selected Problems*. Academic Press Series in Engineering, Elsevier Science, San Diego CA, 2002.

J. H. Keenan. Availability and irreversibility in thermodynamics. *British Journal of Applied Physics*, 2(7): 183–192, 1951.

H. Khalil. *Nonlinear Systems*. Prentice-Hall, Upper Saddle River, NJ, 2002.

N. M. Kirby, L. Xu, M. Luckett, and W. Siepmann. HVDC transmission for large o shore wind farms. *Power Engineering Journal*, 16(3): 135–141, 2002.

P. C. Krause, S. D. Sudhoff, S. Pekarek, and O. Wasynczuk. *Analysis of Electric Machinery and Drive Systems*. IEEE Press, Wiley, 2013.

M. Krstić, P. V. Kokotovic, and I. Kanellakopoulos. *Nonlinear and Adaptive Control Design*. John Wiley & Sons, Inc., New York, 1995.

J. Lee, J. Hong, K. Nam, R. Ortega, A. Astolfi, and L. Praly. Sensorless control of surface-mount permanent magnet synchronous motors based on a nonlinear observer. *IEEE Transactions on Power Electronics*, 25(2): 290–297, 2010.

W. E. Leithead and B. Connor. Control of variable speed wind turbines: design task. *European Journal of Control*, 60(3): 1122–1132, 2013.

W. Leonhard. *Control of Electrical Drives*. Springer-Verlag, Berlin, 1985.

Z. Lu and W. Lin. Asymptotic tracking control of variable-speed wind turbines. *IFAC World Congress*, 18(1): 8457–8462, 2011.

H. Lund. Large-scale integration of wind power into different energy systems. *Energy*, 30(13): 2402–2412, 2005.

B. M. Maschke, R. Ortega, and A. J. van der Schaft. Energy-based Lyapunov functions for forced Hamiltonian systems with dissipation. *IEEE Transactions on Automatic Control*, 45(8): 1498–1502, 2000.

N. Monshizadeh, P. Monshizadeh, R. Ortega, and A. J. van der Schaft. Conditions on shifted passivity of port-Hamiltonian systems. *System & Control Letters*, 123: 55–61, 2019.

J. M. Olm, X. Ros-Oton, and Y. B. Shtessel. Stable inversion of Abel equations: application to tracking control in DC—DC nonminimum phase boost converters. *Automatica*, 47(1): 221–226, 2011.

R. Ortega, J. A. Loría, P. J. Nicklasson, and H. Sira-Ramírez. *Passivity-Based Control of Euler-Lagrange Systems*. Springer-Verlag, 1998.

R. Ortega, N. Monshizadeh, P. Monshizadeh, D. Bazylev, and A. Pyrkin. Permanent magnet synchronous motors are globally asymptotically stabilizable with PI current control. *Automatica*, 98(12): 296–301, 2018.

R. Ortega, A. J. van der Schaft, I. Mareels, and B. M. Maschke. Putting energy back in control. *IEEE Control Systems Magazine*, 21(2): 18–33, 2001.

L. Y. Pao and K. E. Johnson. A tutorial on the dynamics and control of wind turbines and wind farms. In *IEEE American Control Conference*, pages 2076–2089, St. Louis, MO, USA, 2009.

M. Pérez, R. Ortega, and J. R. Espinoza. Passivity-based PI control of switched power converters. *IEEE Transactions on Control Systems Technology*, 12(6): 881–890, 2004.

R. T. Rockafellar. *Convex Analysis*. Princeton University Press, 2015.

J. Salazar, F. Tadeo, K. Witheephanich, M. Hayes, and C. de Prada. *Smart Innovation, Systems and Technologies, chapter Control for a Variable Speed Wind Turbine Equipped with a Permanent Magnet Synchronous Generator (PMSG)*. Springer-Verlag Berlin/Heidelberg, 2012.

S. R. Sanders and G. C. Verghese. Lyapunov-based control for switched power converters. *IEEE Transactions on Power Electronics*, 7(1): 17–24, 1992.

R. Sepulchre, M. Jankovic, and P. Kokotovic. *Constructive nonlinear control*. Springer London Ltd., 2011.

J. W. Simpson-Porco. Equilibrium-independent dissipativity with quadratic supply rates. *IEEE Transactions on Automatic Control*, 64(4): 1440–1455, 2019.

P. Tomei and C. Verrelli. A nonlinear adaptive speed tracking control for sensorless permanent magnet step motors with unknown load torque. *International Journal of Adaptive Control and Signal Processing*, 22: 266–288, 2008.

P. Tomei and C. Verrelli. Observer-based speed tracking control for sensorless PMSMs with unknown load torque. *IEEE Transactions on Automatic Control*, 56(6): 1484–1488, 2011.

S. Trip, M. Bürger, and C. De Persis. An internal model approach to (optimal) frequency regulation in power grids with time-varying voltages. *Automatica*, 64: 240–253, 2016.

F. Valenciaga, P. F. Puleston, and P. E. Battaiotto. Power control of a solar/wind generation system without wind measurement: a passivity/sliding mode approach. *IEEE Transactions on Energy Conversion*, 18(4): 501–507, 2003.

A. J. van der Schaft. *L_2-Gain and Passivity Techniques in Nonlinear Control*. Springer-Verlag, Berlin, 3rd edition, 2016.

J. T. Wen and M. Arcak. A unifying passivity framework for network flow control. *IEEE Transactions on Automatic Control*, 49(2): 162–174, 2004.

A. Yazdani and R. Iravani. *Sourced Controlled Power Converters: Modeling, Control and Applications*. Wiley IEEE, 2010.

D. Zonetti and R. Ortega. *Control of HVDC Transmission Systems: From Theory to Practice and Back*, pages 153–177. Mathematical Control Theory I: Nonlinear and Hybrid Control Systems. Springer-Verlag, Berlin/Heidelberg, 2015.

D. Zonetti, R. Ortega, and A. Benchaib. Modeling and control of HVDC transmission systems from theory to practice and back. *Control Engineering and Practice*, 45(12): 133–146, 2015.

5

Parameterization of All Passive Outputs for Port-Hamiltonian Systems

As discussed in Chapter 2, the main principle underlying PID-PBC is that, PIDs define output strictly passive maps; therefore, if the PID is wrapped around a passive output of the system, $\mathcal{L}_2$-stability of the overall system is guaranteed. Consequently, the identification of the passive outputs of the system to be controlled is a key step in the construction of PID-PBCs. However, the passive output of the system is not unique; therefore, a parameterization of all the possible passive outputs of a system is of utmost importance.

This chapter is devoted to the characterization – in terms of some *free mappings* – of all the passive outputs of a pH system of the form given in (5.1), considering the Hamiltonian $H(x)$ as the storage function. The latter clarification is very important because there might be other passive outputs with respect to storage functions different from the system's Hamiltonian. It is also shown that some particular passive outputs that have been reported in the literature are particular cases of our general result.

5.1 Parameterization of All Passive Outputs

Consider the standard input-state-output representation of a pH system

$$\dot{x} = [\mathcal{J}(x) - \mathcal{R}(x)]\nabla H(x) + g(x)u \tag{5.1a}$$

$$y = h(x) + j(x)u, \tag{5.1b}$$

where $x(t) \in \mathbb{R}^n$ is the state vector, $u(t) \in \mathbb{R}^m$, $m \leq n$ is the control vector, $y(t) \in \mathbb{R}^m$ is an output of the system defined via the mappings $h : \mathbb{R}^n \to \mathbb{R}^m$ and $j : \mathbb{R}^n \to \mathbb{R}^{m\times m}$, $H : \mathbb{R}^n \to \mathbb{R}$ is the Hamiltonian, which we assume

PID Passivity-Based Control of Nonlinear Systems with Applications, First Edition.
Romeo Ortega, José Guadalupe Romero, Pablo Borja, and Alejandro Donaire.

Published 2021 by John Wiley & Sons, Inc.

is bounded from below, $\mathcal{J} : \mathbb{R}^n \to \mathbb{R}^{n\times n}$ and $\mathcal{R} : \mathbb{R}^n \to \mathbb{R}^{n\times n}$, with $\mathcal{J}(x) = -\mathcal{J}^\top(x)$ and $\mathcal{R}(x) = \mathcal{R}^\top(x) \geq 0$, are the interconnection and damping matrices, respectively, and $g : \mathbb{R}^n \to \mathbb{R}^{n\times m}$ is the input matrix, which is full rank. The first step toward the parameterization of the passive outputs of a pH system consists in computing a nonunique factorization of the dissipation matrix of the form

$$\mathcal{R}(x) = \phi^\top(x)\phi(x), \tag{5.2}$$

where $\phi : \mathbb{R}^n \to \mathbb{R}^{q\times n}$, with $q \in \mathbb{N}$ satisfying

$$q \geq \text{rank}\,\{\mathcal{R}(x)\}.$$

We recall the basic linear algebra fact that $\mathcal{R}(x) \geq 0$ *if and only if* such a factor exists (Horn and Johnson, 1985).

Proposition 5.1: *Consider the pH system (5.1). The following statements are equivalent:*

S1 *The mapping $\Sigma : u \mapsto y$ is passive with storage function $H(x)$.*
S2 *For any factorization of the dissipation matrix $\mathcal{R}(x)$ of the form (5.2) the mappings $h(x)$ and $j(x)$ can be expressed as*

$$\begin{aligned} h(x) &= [g(x) + 2\phi^\top(x)w(x)]^\top \nabla H(x) \\ j(x) &= w^\top(x)w(x) + D(x), \end{aligned} \tag{5.3}$$

for some mappings $w : \mathbb{R}^n \to \mathbb{R}^{q\times m}$ and $D : \mathbb{R}^n \to \mathbb{R}^{m\times m}$, with $D(x) = -D^T(x)$.

Proof. It is well-known, see Proposition A.1 in Appendix A, that the system (5.1) is passive if and only if

$$\begin{bmatrix} -2[\nabla H(x)]^\top \mathcal{R}(x)\nabla H(x) & [\nabla H(x)]^\top g(x) - h^\top(x) \\ g^\top(x)\nabla H(x) - h(x) & -[j(x) + j^\top(x)] \end{bmatrix} \leq 0.$$

To prove that **S2** implies **S1** substitute (5.2) and the definition of $h(x)$ and $j(x)$ of (5.1b) above to get

$$2\begin{bmatrix} -|\phi(x)\nabla H(x)|^2 & -[\nabla H(x)]^\top \phi^\top(x)w(x) \\ -w^\top(x)\phi(x)\nabla H(x) & -w^\top(x)w(x) \end{bmatrix} \leq 0,$$

which is always satisfied.

The proof that **S1** implies **S2** proceeds following (Sepulchre et al., 2011). Namely, assume $u \mapsto y$ is passive with storage function $H(x)$ and define the mapping $d : \mathbb{R}^n \times \mathbb{R}^m \to \mathbb{R}_+$

$$d(x,u) := -\dot{H} + u^\top[h(x) + j(x)u] \geq 0. \tag{5.4}$$

Evaluating $\dot{H}$ along the trajectories of (5.1a) and using (5.2), we get

$$d(x,u) = |\phi(x)\nabla H(x)|^2 + u^\top[h(x) - g^\top(x)\nabla H(x)] + \frac{1}{2}u^\top[j(x) + j^\top(x)]u.$$

Because $d(x,u)$ is quadratic in $u(t)$ and nonnegative for all $u(t)$ and $x(t)$, there exists a (nonunique) matrix valued function $w(x)$ such that

$$d(x,u) = |\phi(x)\nabla H(x) + w(x)u|^2.$$

The proof that $h(x)$ and $j(x)$ take the form (5.3) is established equating the terms of like power in u and invoking the skew-symmetry property of $D(x)$. □

We now denote the passive output (5.1b) by y_{wD} to indicate that $h(x)$ and $j(x)$ are parameterized as in (5.3).

5.2 Some Particular Cases

In this section, we prove that all passive outputs of the pH system (5.1) reported in the literature can be generated via the parameterization of the passive output provided in (5.3). To that end, we introduce the following proposition whose proof may be found in Zhang et al. (2018):

Proposition 5.2: *Consider the output (5.1b) and its parameterization (5.3). The following implications hold true:*

- *Natural output (Ortega et al., 2001; van der Schaft, 2016):*

$$\left.\begin{array}{l} w(x) = 0 \\ D(x) = 0 \end{array}\right\} \Rightarrow y_{wD} = y_0 = g^\top(x)\nabla H(x). \tag{5.5}$$

- *Power-shaping output of (Ortega et al., 2003) with $F(x)$ full rank:*[1]

$$\left.\begin{array}{l} w(x) = \phi(x)F^{-1}(x)g(x) \\ D(x) = -g^\top(x)F^{-\top}(x)\mathcal{J}(x)F^{-1}(x)g(x) \end{array}\right\} \Rightarrow y_{wD} = -g^\top(x)F^{-\top}(x)\dot{x}. \tag{5.6}$$

- *The alternate output of (Duindam et al., 2009; Ortega et al., 2007, Venkatraman and van der Schaft, 2010) with generalized damping matrix verifying*

$$\mathcal{Z}(x) := \begin{bmatrix} \mathcal{R}(x) & T(x) \\ T^\top(x) & S(x) \end{bmatrix} \geq 0 : \tag{5.7}$$

1 We recall that $F(x) := \mathcal{J}(x) - \mathcal{R}(x)$.

$$\left.\begin{aligned} S(x) &= w^\top(x)w(x) \\ T(x) &= \phi^\top(x)w(x) \end{aligned}\right\} \Rightarrow\ y_{wD} = [g(x) + 2T(x)]^\top \nabla H(x) + [S(x) + D(x)]u. \tag{5.8}$$

- *Power-shaping output of (Ortega et al., 2003) with* $F(x)$ not full rank *but satisfying*

$$F^\top(x)[F^\dagger(x)]^\top(x)F(x) = F(x) \tag{5.9a}$$

$$\text{span}\{g(x)\} \subseteq \text{span}\{F(x)\}, \tag{5.9b}$$

where $F^\dagger(x)$ *is a pseudo-inverse of* $F(x)$*:*

$$\left.\begin{aligned} w(x) &= \phi(x)F^\dagger(x)g(x) \\ D(x) &= -g^\top(x)[F^\dagger(x)]^\top(x)\mathcal{J}(x)F^\dagger(x)g(x) \end{aligned}\right\} \Rightarrow\ y_{wD} = -g^\top(x)[F^\dagger(x)]^\top \dot{x}. \tag{5.10}$$

5.3 Two Additional Remarks

R1 In Proposition 5.1, the storage function $H(x)$ and the matrix $F(x)$ are *fixed*, and passivity is established with respect to the storage function $H(x)$ and that particular matrix $F(x)$. It is clear that the pH representation is not unique and there might be other factorizations for the vector field $F(x)\nabla H(x)$. In this way, new passive outputs – with different storage functions – can be generated. This fact is illustrated in one of the examples provided in Section 5.4.

R2 Sometimes it is possible to identify, for a given system, different passive outputs with their corresponding storage functions. That is, when there exist $r \in \mathbb{N}$ storage functions $H_i : \mathbb{R}^n \to \mathbb{R}$, and output signals $y_i(t) \in \mathbb{R}^m$, such that

$$\dot{H}_i = u^T y_i - d_i,\ i = 1, 2, \dots, r$$

where $d_i : \mathbb{R}^n \to \mathbb{R}_{\geq 0}$ are the dissipation functions. It is clear that the mapping $u \mapsto \sum_{i=1}^{r} a_i y_i$, with $a_i \geq 0$, is passive with storage function $\sum_{i=1}^{r} a_i H_i(x)$.

As discussed in Chapter 7 – see also Donaire et al. (2016), Romero et al. (2016), and Romero et al. (2018) – to shape the energy of mechanical systems with PID-PBCs, it is sometimes necessary to *change the sign* of some terms of the energy function. Think, for instance, of the problem of stabilization of the upward (zero) position of the hanging pendulum, whose potential energy function is a cosine function that has a maximum, instead of a minimum, at the desired position. In this case, the potential energy function is flipped up by multiplying it by a *negative* coefficient a_i. If there is no damping, the new output may be still a passive output provided the new storage function is positive definite. See the discussion in Remark 7.2.

5.4 Examples

In this section, we illustrate the results presented in the chapter with two physical examples.

5.4.1 A Level Control System

Consider the two-tanks system shown in Figure 5.1, where the state variables $x_i(t) > 0$, $i = 1, 2$, represent the water level in the corresponding tank, $u(t)$ is the flow pumped from the reservoir, and $b \in (0, 1]$ is a constant parameter that indicates the position of the valve, where $b = 1$ means that the valve is fully open. This system admits a pH representation of the form (5.1a) with

$$\mathcal{J}(x) = \begin{bmatrix} 0 & \alpha_2\sqrt{x_2} \\ -\alpha_2\sqrt{x_2} & 0 \end{bmatrix}, \quad \mathcal{R}(x) = \begin{bmatrix} \alpha_1\sqrt{x_1} & 0 \\ 0 & 0 \end{bmatrix}, \quad g = \begin{bmatrix} 1 \\ \frac{1-b}{b} \end{bmatrix} \tag{5.11}$$

and Hamiltonian

$$H(x) = x_1 + \frac{\alpha_1}{\alpha_2}x_2, \tag{5.12}$$

where $\alpha_i > 0$ are constant parameters. Accordingly, the natural passive output of this system is given by

$$y_0 = 1 + \frac{\alpha_1}{\alpha_2}\frac{1-b}{b}. \tag{5.13}$$

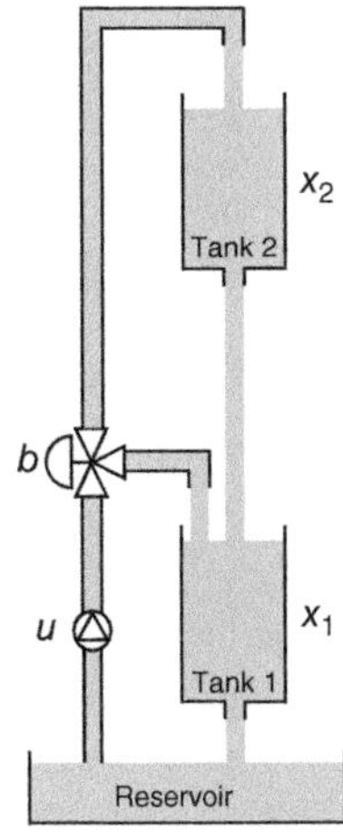

Figure 5.1 Two-tanks system.

However, as it is shown in Ortega et al. (2007), this system admits an alternative pH representation of the form $F(x)\nabla H(x) = \overline{F}(x)\nabla\overline{H}(x)$, with

$$\overline{F}(x) = -\frac{1}{ad}\begin{bmatrix} d & -a \\ 0 & a \end{bmatrix}, \quad \overline{H}(x) = \tfrac{2}{3}a\alpha_1 x_1^{\frac{3}{2}} + \tfrac{2}{3}d\alpha_2 x_2^{\frac{3}{2}},$$

where $4d > a > 0$. Therefore, a passive output for this system is given by

$$\overline{y} = a\alpha_1\sqrt{x_1} + \frac{1-b}{b}d\alpha_2\sqrt{x_2}, \tag{5.14}$$

where some simple computations show that $\overline{y}$, given in (5.14), cannot be obtained via the parameterization (5.3) if we consider $H(x)$ as the storage function of the system.

5.4.2 A Microelectromechanical Optical Switch

Consider a microelectromechanical optical switch which admits a pH representation of the form (5.1), with

$$F(x) = \begin{bmatrix} 0 & 1 & 0 \\ -1 & -b & 0 \\ 0 & 0 & -\frac{1}{r} \end{bmatrix}, \quad g = \begin{bmatrix} 0 \\ 0 \\ \frac{1}{r} \end{bmatrix} \tag{5.15}$$

$$H(x) = \frac{1}{2}ax_1^2 + \frac{1}{2m}x_2^2 + \frac{1}{2c_1(x_1 + c_0)}x_3^2,$$

where x_1 and x_2 denote the position of the comb driver actuator and its momentum, respectively, x_3 represents the charge on the capacitor, u is the voltage applied to the electrodes, and $a, b, r, m, c_0, c_1 > 0$ are constant parameters. This system is subject to the physical constraint $x_1 \geq 0$.

Consider the parameterization of the passive output provided in (5.3). Hence, by fixing the parameters $w(x)$ and $D(x)$ as zero, we get that

$$y_{\mathrm{wD}} = y_0 = g^\top \nabla H(x) = \frac{x_3}{rc_1(x_1 + c_0)}.$$

On the other hand, since $F(x)$ has full rank, by selecting $w(x)$ and $D(x)$ as in (5.6), we obtain

$$y_{\mathrm{wD}} = -g^\top F^{-\top}\dot{x} = \dot{x}_3.$$

Therefore, it is clear that different selections of the free parameters $w(x)$ and $D(x)$ lead to different passive outputs, where one of the main motivations of considering one passive output over another is its integrability property, which, as is discussed in Chapter 6, is essential in the design of PID-PBCs.

Bibliography

A. Donaire, R. Mehra, R. Ortega, S. Satpute, J. G. Romero, F. Kazi, and N. M. Singh. Shaping the energy of mechanical systems without solving partial differential equations. *IEEE Transactions on Automatic Control*, 61(4): 1051–1056, 2016.

V. Duindam, A. Macchelli, S. Stramigioli, and H. Bruyninckx. *Modeling and Control of Complex Physical Systems: The Port-Hamiltonian Approach*. Springer Science & Business Media, 2009.

R. Horn and C. Johnson. *Matrix Analysis*. Cambridge University Press, 1985.

R. Ortega, D. Jeltsema, and J. M. A. Scherpen. Power shaping: a new paradigm for stabilization of nonlinear RLC circuits. *IEEE Transactions on Automatic Control*, 48(10): 1762–1767, 2003.

R. Ortega, A. J. van der Schaft, F. Castaños, and A. Astolfi. Control by state-modulated interconnection of port-Hamiltonian systems. In *IFAC Symposium on Nonlinear Control Systems*, Pretoria, South Africa, 2007.

R. Ortega, A. J. van der Schaft, I. Mareels, and B. M. Maschke. Putting energy back in control. *IEEE Control Systems Magazine*, 21(2): 18–33, 2001.

J. G. Romero, A. Donaire, R. Ortega, and P. Borja. Global stabilisation of underactuated mechanical systems via PID passivity-based control. *Automatica*, 96(10): 178–185, 2018.

J. G. Romero, R. Ortega, and A. Donaire. Energy shaping of mechanical systems via PID control and extension to constant speed tracking. *IEEE Transactions on Automatic Control*, 61(11): 3551–3556, 2016.

R. Sepulchre, M. Jankovic, and P. Kokotovic. *Constructive nonlinear control*. Springer London Ltd., 2011.

A. J. van der Schaft. *L_2-Gain and Passivity Techniques in Nonlinear Control*. Springer-Verlag, Berlin, 3rd edition, 2016.

A. Venkatraman and A. J. van der Schaft. Energy shaping of port-Hamiltonian systems by using alternate passive input-output pairs. *European Journal of Control*, 16(6): 665–677, 2010.

M. Zhang, P. Borja, R. Ortega, Z. Liu, and H. Su. PID passivity-based control of port-Hamiltonian systems. *IEEE Transactions on Automatic Control*, 63(4): 1032–1044, 2018.

6

Lyapunov Stabilization of Port-Hamiltonian Systems

As explained in Chapter 2, the central idea of PID-PBC is to exploit the output strict passivity property of the operator Σ_c defined by the PID (2.1) to, putting it in feedback with a passive system, ensure the $\mathcal{L}_2$-stability of the overall system. Toward this end, the parameterization of all passive outputs for pH systems, considering the Hamiltonian of the system as the storage function, provided in Chapter 5, plays a key role.

Although important, $\mathcal{L}_2$-stability is a rather weak property that does not even guarantee boundedness in the presence of external disturbances, see Kellett and Dower (2016) for a detailed discussion. In this chapter, we investigate under which conditions it is possible to use PID-PBC to achieve the more ambitious objective of *Lyapunov stability* of a desired equilibrium point for pH systems. More precisely, we consider the standard input-state-output representation of a pH system given in (5.1), with y_{wD} the corresponding passive output with storage function $H(x)$, and a point $x^\star$ element of the set of assignable equilibria (B.2). See Section B.2 for the construction of $\mathcal{E}$. The *control objective* is to design a PID-PBC of the form

$$
\begin{aligned}
u &= -K_P y_{\mathrm{wD}} - K_I x_c - K_D \dot{y}_{\mathrm{wD}}, \\
\dot{x}_c &= y_{\mathrm{wD}},
\end{aligned}
\tag{6.1}
$$

such that $(x^\star, x_c^\star)$ is a stable equilibrium point for the closed-loop system.[1]

As pointed out in Corollary 4.2.2 of (van der Schaft, 2016a), a passive system is Lyapunov stable if the storage function has a strict minimum at the equilibrium point – moreover, it is asymptotically stable if the passive output is detectable. Satisfying these two requirements is the central

1 As is shown in Appendix B, for any $x^\star \in \mathcal{E}$ there exists $u^\star$ given in (B.4). Therefore, from (6.1), it follows that for any $x^\star \in \mathcal{E}$, there exists $x_c^\star$ given by $x_c^\star = K_I^{-1} u^\star$.

PID Passivity-Based Control of Nonlinear Systems with Applications, First Edition.
Romeo Ortega, José Guadalupe Romero, Pablo Borja, and Alejandro Donaire.

Published 2021 by John Wiley & Sons, Inc.

objective of this chapter. To comply with the positive definite requirement of the closed-loop storage function, we propose to generate a Lyapunov function depending *only* on the system state, whose construction proceeds as follows. First, we recall from Chapter 2 that the storage function of the closed-loop system is given by

$$U(x, x_c) = H(x) + \frac{1}{2}\|h(x)\|_{K_D} + \frac{1}{2}\|x_c\|_{K_I},$$

where $x(t)$ and $x_c(t)$ are the states of the plant and the controller, respectively, which is a nonincreasing function. The objective is to construct *invariant* sets of the form

$$\mathcal{M}_\kappa := \{(x, x_c) \in \mathbb{R}^n \times \mathbb{R}^m \mid x_c = \gamma(x) + \kappa\}, \tag{6.2}$$

where $\gamma : \mathbb{R}^n \to \mathbb{R}^m$ is a mapping to be defined and $\kappa \in \mathbb{R}$ – see Definition C.1 for the definition of invariant sets and the necessary and sufficient conditions for their existence. In this way, it is possible to define a function

$$H_d(x) := H(x) + \frac{1}{2}\|h(x)\|_{K_D} + \frac{1}{2}\|\gamma(x) + \kappa\|_{K_I}, \tag{6.3}$$

that, in $\mathcal{M}_\kappa$, satisfies

$$U(x(t), x_c(t)) = H_d(x(t)), \tag{6.4}$$

hence, is *nonincreasing*. If we are able to prove that $H_d(x)$ is positive definite (with respect to $x^\star$) then it qualifies as Lyapunov function – achieving our objective. Since $\mathcal{M}_\kappa$ is invariant, it remains to ensure that the desired equilibrium is in this set and that the system's trajectory "starts" also there, an issue that is resolved below.

After identifying the PDE that needs to be solved and presenting the Lyapunov stability analysis, we discuss in the chapter methods to *solve* the PDE explicitly. Then, we propose an alternative procedure to *add the derivative action*. This is motivated by two facts, on the one hand, that this can be implemented – without differentiation – only for systems with relative degree greater than zero. On the other hand, as discussed in Proposition 2.2, the dissipation obstacle, in this case, hampers the possibility to assign an equilibrium to the closed loop. We wrap up the chapter with a series of illustrative examples.

6.1 Generation of Lyapunov Functions

In this section, we propose a procedure to construct the Lyapunov function (6.3) discussed above, whose key step is the identification of the invariant sets $\mathcal{M}_\kappa$ (6.2). Following Lemma C.1, this entails the solution of a PDE, in this case, associated with the condition:

$$\dot{x}_c = \frac{d}{dt}\gamma(x), \tag{6.5}$$

which is the necessary and sufficient condition for the invariance of the sets $\mathcal{M}_\kappa$. The strategy that we follow to satisfy (6.5) is to look for the *free* mappings $w(x)$ and $D(x)$, that define the passive output $y_{\rm wD}$, see Proposition 5.1.

In the sequel, we first identify the PDE associated with (6.5), then we give additional conditions that ensure it is a positive definite function.

6.1.1 Basic PDE

The main result of the section is summarized in the proposition below.

Proposition 6.1: *Consider the pH system* (5.1) *with $h(x)$ and $j(x)$ defined in* (5.3) *and $\dot{x}_c = y_{wD}$. Assume there exist mappings $w(x)$ and $D(x) = -D^\top(x)$ such that the PDE*

$$\begin{bmatrix} [\nabla H(x)]^\top F^\top(x) \\ g^\top(x) \end{bmatrix} \nabla\gamma(x) = \begin{bmatrix} [\nabla H(x)]^\top [g(x) + 2\phi^\top(x) w(x)] \\ w^\top(x) w(x) - D(x) \end{bmatrix} \tag{6.6}$$

admits a solution $\gamma : \mathbb{R}^n \to \mathbb{R}^m$. Then, the set $\mathcal{M}_\kappa$ is invariant. Moreover, the projected energy function (6.3) *with*

$$\kappa := x_c(0) - \gamma(x(0))$$

verifies the identity (6.4) *for all $t \geq 0$.*

Proof. The proof is established verifying (6.5) and noting that $y_{\rm wD} = \dot{x}_c$. Substituting (5.1) and (5.3) in the latter identity yields

$$\begin{aligned}[\nabla\gamma(x)]^\top [F(x)\nabla H(x) + g(x)u] &= [g(x) + 2\phi^\top(x) w(x)]^\top \nabla H(x) \\ &\quad + [w^\top(x) w(x) + D(x)]u.\end{aligned}$$

The proof of invariance of $\mathcal{M}_\kappa$ is completed equating the terms dependent and independent on u, respectively. The second claim follows integrating (6.5). □

Remark 6.1: As discussed in Castaños et al. (2009) and Zhang et al. (2018) a drawback of the construction described above is that, to ensure the state (x, x_c) lives in the foliation of $\mathcal{M}_\kappa$ that contains the desired equilibrium, it is necessary to fix the initial conditions of x_c, see Eq. (42) of (Zhang et al., 2018). This makes the stability analysis "trajectory-dependent," hence, intrinsically fragile – see the discussion on this matter in Ortega (2021). To avoid this problem, it is suggested in Proposition 8 of (Zhang et al., 2018) to project onto $\mathcal{M}_\kappa$ also the PID, and implement it as a static-state feedback. The details of this construction are given in Section 6.1.2.

6.1.2 Lyapunov Stability Analysis

In this section, we complete the Lyapunov stability analysis by ensuring the projected function $H_d(x)$ is positive definite. To simplify the presentation of the material, we avoid the technical details pertaining to the required well-posedness conditions of the feedback loop and refer the reader to Lemma 2.2 for the details. The proof of the proposition is established invoking standard Lyapunov stability theory (Khalil, 2002).

Proposition 6.2: *Consider the pH system* (5.1) *with passive output* $y_{wD} = h(x)$ *such that the PDE* (6.6) *admits a solution* $\gamma(x)$. *Fix an equilibrium* $x^\star \in \mathcal{E}$ *and define the* projected *PID-PBC as*

$$u = -K_P y_{\text{wD}} - K_I[\gamma(x) - \kappa] - K_D \dot{y}_{\text{wD}}, \tag{6.7}$$

with

$$\kappa := -(\gamma^\star + K_I^{-1} u^\star), \tag{6.8}$$

and $u^\star$ *given in (B.4). Assume that* $H_d(x)$, *given in* (6.3), *satisfies*

$$x^\star = \arg\min H_d(x), \tag{6.9}$$

and $x^\star$ *is isolated.*

(i) The closed-loop system has a stable *equilibrium at* $x^\star$ *with Lyapunov function* (6.3).
(ii) The equilibrium is asymptotically *stable if the signal* y_{wD} *is a* detectable *output for the closed-loop system.*
(iii) The stability properties are global *if* $H_d(x)$ *is radially unbounded.*

Proof. First, we need to prove that $x^\star$ is an equilibrium of the system (5.1) in closed-loop with (6.7). To this end, note that $y_{\text{wD}} = \dot{\gamma}$ implies that $y_{\text{wD}} = 0$ at the equilibrium that, replaced in (6.7), yields u equals $u^\star$. This part of the proof is completed invoking the equivalence (B.4).

Now, from (6.3), we have that

$$\begin{aligned} \dot{H}_d &\leq y_{\text{wD}}^\top u + h^\top(x) K_D \dot{h} + [\gamma(x) + \kappa]^\top K_I \dot{\gamma} \\ &= -y_{\text{wD}}^\top [K_P y_{\text{wD}} + K_I(\gamma(x) + \kappa) + K_D \dot{y}_{\text{wD}}] + y_{\text{wD}}^\top [K_I(\gamma(x) + \kappa) + K_D \dot{y}_{\text{wD}}] \\ &= -\|y_{\text{wD}}\|_{K_P}^2 \leq 0, \end{aligned} \tag{6.10}$$

where we used the passivity property $\dot{H} \leq y_{\mathrm{wD}}^{\top} u$ and $y_{\mathrm{wD}} = h(x) = \dot{\gamma}$. From (6.10), it follows that $H_d(x(t))$ is a nonincreasing function that, moreover, is positive definite because of the assumption (6.9). The proof is completed invoking standard Lyapunov stability theory (Khalil, 2002). □

Remark 6.2: Notice that the second right-hand term in (6.7) is an *effective integral action*. Indeed, in view of (6.5), the controller may be rewritten as (6.1) with $x_c(0) = \kappa + \gamma(x(0))$.

Remark 6.3: Note that the derivative action can be implemented (without differentiation) *only* if $j(x) = 0$. Hence, if this is not the case, $K_D = 0$ in (6.7).

6.2 Explicit Solution of the PDE

As is well known, finding a solution to the PDE (6.6) is, in general, a challenging task. In this section, we discuss several methods for its explicit solution and propose the use of multiplier theory (Desoer and Vidyasagar, 2009) to extend the realm of applicability of the theory.

6.2.1 The Power Shaping Output

The main technical tool that we use in this section is Poincare's lemma, which gives necessary and sufficient conditions for a vector field to be the gradient of a scalar function, called *gradient vector fields* – see Section C.2. To see the connection with the problem at hand, recall that we are trying to find an output y_{wD} such that $y_{\mathrm{wD}} = \frac{d}{dt}\gamma(x)$, for some mapping $\gamma(x)$. This is clearly equivalent to

$$y_{\mathrm{wD}} = [\nabla\gamma(x)]^{\top}\dot{x} =: \beta^{\top}(x)\dot{x}, \tag{6.11}$$

where we have *postulated* the existence of a new mapping $\beta : \mathbb{R}^n \to \mathbb{R}^{n\times m}$. From Poincare's lemma, we know that this mapping exists if and only if its m columns $\beta_i : \mathbb{R}^n \to \mathbb{R}^n$ are *integrable*, that is they satisfy

$$\nabla\beta_i(x) = \left[\nabla\beta_i(x)\right]^{\top}. \tag{6.12}$$

In that case, we can explicitly compute $\gamma(x)$ via the simple integration:

$$\gamma(x) = \int_0^1 \beta^{\top}(sx)x\, ds + \gamma(0). \tag{6.13}$$

As shown in Proposition 5.2, the power shaping outputs are precisely of the form (6.11) for suitably defined maps $\beta(x)$. Imposing the integrability condition (6.12) to this mappings give us then, explicit solutions to the PDE. The result is summarized in the proposition below whose proof may be found in Zhang et al. (2018).

Proposition 6.3: *Consider the pH system* (5.1).

- *Assume the matrix $F(x)$ is full rank and the vectors $F^{-1}(x)g_i(x) \in \mathbb{R}^n$, $i = 1, \ldots, m$, are gradient vector fields. Then,*

$$\gamma(x) = -\int_0^1 [F^{-1}(sx)g(sx)]^\top x \, ds + \gamma(0)$$

is a solution of the PDE (6.6) *with $w(x)$ and $D(x)$ given in* (5.6).
- *Assume the matrix $F(x)$ is* not *full rank, but satisfies* (5.9), *and the vectors $F^\dagger(x)g_i(x) \in \mathbb{R}^n$, $i = 1, \ldots, m$, are gradient vector fields. Then,*

$$\gamma(x) = -\int_0^1 [F^\dagger(sx)g(sx)]^\top x \, ds + \gamma(0),$$

is a solution of (6.6) *with the parameters $w(x)$ and $D(x)$ given in* (5.10).

Remark 6.4: Derivations similar to the ones done in Proposition 6.3 are reported in Section 7.1 of (van der Schaft, 2016) where, following the construction of (Maschke et al., 2000), new passive outputs – called "alternate" in van der Schaft (2016) – are used for CbI. There is a relation also with input–output Hamiltonian systems with dissipation studied in van der Schaft (2016b), for which the integrability condition (6.12) is implicitly assumed. See these references for further details.

6.2.2 A More General Solution

The construction given in Proposition 6.3 pertains to the solution of the PDE

$$\begin{bmatrix} F^\top(x) \\ g^\top(x) \end{bmatrix} \nabla\gamma(x) = \begin{bmatrix} g(x) + 2\phi^\top(x)w(x) \\ w^\top(x)w(x) - D(x) \end{bmatrix}. \tag{6.14}$$

As discussed in Chapter 2, this is the PDE that is solved in CbI, where the energy function is shaped generating Casimir functions for the interconnected pH system. Comparing (6.6) with (6.14), we note the absence of the term $\nabla H(x)$ in the first set of equations, which is due to the fact that Casimir functions are independent of $H(x)$.

In this section, we propose to exploit the presence of the term $\nabla H(x)$ to enlarge the set of solution of the PDE (6.6). Toward this end, we introduce the following:

Assumption 6.1: Given $x^\star \in \mathcal{E}$, there exist mappings $\alpha(x) \in \mathbb{R}^{n\times m}$ and $\rho(x) \in \mathbb{R}^{q\times m}$ such that

$$\alpha^\top(x)\nabla H(x) = 0, \tag{6.15}$$

$$\alpha^\star = 0, \tag{6.16}$$

$$\mathtt{sym}\left\{\alpha^\top(x)F^{-1}(x)g(x)\right\} = \rho^\top(x)\rho(x). \tag{6.17}$$

Moreover, the columns of the matrix

$$\beta(x) := F^{-\top}(x)[\alpha(x) + 2\phi^\top(x)\rho(x)] - F^{-1}(x)g(x), \tag{6.18}$$

are gradient vector fields.

Equipped with this assumption, we can state the main result of the section whose proof may be found in Borja et al. (2020).

Proposition 6.4: *If Assumption 6.1 holds, the following definitions of $w(x)$ and $D(x)$:*

$$\begin{aligned} w(x) &:= \rho(x) + \phi(x)F^{-1}(x)g(x) \\ D(x) &:= -g^\top(x)F^{-\top}(x)\mathcal{J}(x)F^{-1}(x)g(x) \\ &\quad + \mathtt{skew}\{[\alpha^\top(x) + 2\rho^\top(x)\phi(x)]F^{-1}(x)g(x)\} \end{aligned}$$

are such that

$$\gamma(x) = \int_0^1 \beta^\top(sx)x\, ds + \gamma(0),$$

is a solution to the PDE (6.6).

Remark 6.5: The explicit solution to the PDE given in the proposition above contains, as *particular case*, the one given in Proposition 6.3. Indeed, the latter is obtained by simply setting $\alpha(x) = 0$ – consequently, $\rho(x) = 0$ – in (6.18) to get $\beta(x) = -F^{-1}(x)g(x)$.

6.2.3 On the Use of Multipliers

Proposition 6.3 establishes algebraic conditions to verify the integrability of y_{wD}. However, such conditions turn out to be only sufficient. Indeed, if the results of the mentioned proposition are not suitable to prove the integrability of the passive output, the use of *multipliers* may offer an alternative to ensure that the passive output is *integrable* without the necessity of solving the PDE (6.6). Toward this end, we introduce a *full rank* matrix $M : \mathbb{R}^n \to \mathbb{R}^{m\times m}$ and the new input–output pair:

$$\bar{u} := M^{-1}(x)u, \quad \bar{y} := M^{\top}(x)y_{\mathrm{wD}}.$$

Notice that the power balance inequality is preserved for these new port variables, i.e.,

$$\dot{H} \leq u^{\top} y_{\mathrm{wD}} = \bar{u}^{\top}\bar{y}. \tag{6.19}$$

Hence, the integrability condition imposed by the PDE 6.1 is now subject to the existence of a mapping $\gamma : \mathbb{R}^n \to \mathbb{R}^m$ such that

$$\bar{y} = \dot{\gamma}. \tag{6.20}$$

Proposition 6.5 establishes some conditions to ensure the existence of a full-rank matrix $M(x)$ such that (6.20) is satisfied.

Proposition 6.5: *Consider the pH system* (5.1), *satisfying* rank $\{F\} = n$, *with passive output*

$$y_{wD} = \beta^{\top}(x)\dot{x},$$

where $\beta(x)$ *is defined as in* (6.18). *Define a mapping* $\Lambda : \mathbb{R}^n \to \mathbb{R}^{n\times(n-m)}$ *satisfying*

(C1)

$$\operatorname{rank}\{\Lambda(x)\} = n - m.$$

(C2)

$$\beta^{\top}(x)\Lambda(x) = 0. \tag{6.21}$$

There exists a full rank matrix $M : \mathbb{R}^n \to \mathbb{R}^{m\times m}$ *such that*

$$\beta(x)M(x) = \nabla\gamma(x), \tag{6.22}$$

where $\gamma : \mathbb{R}^n \to \mathbb{R}^m$, if and only if *for all* $1 \leq i,j \leq n - m$

$$\operatorname{rank}\left\{\left[\Lambda(x) \;\vdots\; [\Lambda_i(x), \Lambda_j(x)]\right]\right\} = n - m, \tag{6.23}$$

where $\Lambda_i(x)$ *is the ith column of* $\Lambda(x)$ *and* $[\cdot,\cdot]$ *is the standard Lie bracket (Spivak, 1999).*

Proof. The proof proceeds as follows: first, recall Frobenius theorem, see Vidyasagar (1993). Given the $n - m$ linearly independent vectors $\Lambda_i(x)$, there exist functions $\gamma_i : \mathbb{R}^n \to \mathbb{R}$ such that

(i) the vectors $\nabla\gamma_k(x)$ are linearly independent, and
(ii) $[\nabla\gamma_k(x)]^\top \Lambda_j(x) = 0,\ k = 1, \dots, m,\ j = 1, \dots, m,$ if and only if (6.23) is satisfied.

The proof is completed noting that, since $M(x)$ is full rank, we have

$$\text{Ker}\left\{M^\top(x)\beta^\top(x)\right\} = \text{Ker}\left\{\beta^\top(x)\right\}, \tag{6.24}$$

and recalling, from (6.21), that the columns of $\Lambda(x)$ are a basis for this space.
□

6.3 Derivative Action on Relative Degree Zero Outputs

Due to its "prediction-like" feature, adding a derivative action to the controller is essential in some applications. As indicated in Remark 6.3, this is, unfortunately, possible only for relative degree one outputs. This section focuses on the implementation of a derivative term for relative degree zero outputs, which are of central importance for the following reason: we recall that the design of the Lyapunov stabilizing PID-PBC relies on the selection of mappings $w(x)$ and $D(x)$ to solve the PDE (6.6). According to Proposition 5.1, these mappings have to be set to zero if we impose the restriction of $j(x) = 0$, losing in this way these degrees of freedom.

To solve the problem mentioned above, we proceed in three steps:

S1 Inject an integral term with y_{wD}.
S2 Identify a *new passive output*, which is now of relative degree one.
S3 Complete the design with the addition of the proportional and derivative terms.

6.3.1 Preservation of the Port-Hamiltonian Structure of I-PBC

The first step in the new design is to prove that the pH structure is preserved after the injection of the integral term with the output y_{wD}. Although this fact is well known for the natural output (Ortega and García-Canseco, 2004, Proposition 4), the identification of the pH structure of the closed-loop system for the general y_{wD} requires a more elaborated analysis which is provided in Proposition 6.6, whose proof may be found in Borja et al. (2020).

This result plays a key role in the construction of the new PID-PBC for systems with relative degree zero. Indeed, the preservation of the pH structure naturally leads to the definition of a *new passive output*, which is defined in the extended state space (x, x_c). Now, using the solution of the PDE (6.6), it is possible to project the dynamics onto the invariant manifold $\mathcal{M}_\kappa$ to define a new passive output – function of the plant's state x only. As shown in Proposition 6.8, the derivative of this passive output can be computed, without differentiation, to add a derivative term to the controller.

Proposition 6.6: *Consider the pH system (5.1) in closed-loop with (6.1). Fix $K_D = 0$ and $K_P = 0$. The closed-loop dynamics, in the augmented state space $\mathrm{col}(x, x_c) \in \mathbb{R}^{n+m}$, has the pH structure*

$$\begin{bmatrix} \dot{x} \\ \dot{x}_c \end{bmatrix} = \mathcal{F}(x)\nabla U_I(x, x_c) + \mathcal{G}(x)v,$$

with storage function

$$U_I(x, x_c) = H(x) + \frac{1}{2}\|x_c\|^2_{K_I},$$

and

$$\mathcal{F}(x) \ := \ \begin{bmatrix} F(x) & -g(x) \\ g^\top(x) + 2w^\top(x)\phi(x) & -j(x) \end{bmatrix}, \quad \mathcal{G}(x) := \begin{bmatrix} g(x) \\ j(x) \end{bmatrix},$$

where

$$\mathtt{sym}\{\mathcal{F}(x)\} = -\begin{bmatrix} \phi^\top(x)\phi(x) & -\phi^\top(x)w(x) \\ -w^\top(x)\phi(x) & w^\top(x)w(x) \end{bmatrix} \leq 0.$$

Moreover,

$$\dot{U}_I = -|\phi(x)\nabla H(x) - w(x)K_I x_c|^2 + v^\top y_I,$$

where

$$y_I := g^\top(x)\nabla H(x) + j^\top(x)K_I x_c.$$

Hence, the map $v \to y_\mathrm{I}$ is passive.

6.3.2 Projection of the New Passive Output

In this section, we project onto the manifold $\mathcal{M}_\kappa$ (6.2) the dynamics of the system (5.1) in closed-loop with an integral action to define a new passive output that depends exclusively on x – the proof of this result may be found in Borja et al. (2020). One important motivation to consider this new property is that, as shown in Proposition 6.8, with this output, it is possible to add a derivative term to our controller.

Proposition 6.7: *Consider the system (5.1), satisfying Assumption 6.1, in closed-loop with the integral controller*

$$u = -K_I[\gamma(x) + \kappa] + v. \tag{6.25}$$

Define[2]

$$y_{\mathrm{IP}} := g^\top(x)\nabla U_{\mathrm{IP}}(x), \tag{6.26}$$

with

$$U_{\mathrm{IP}}(x) := H(x) + \frac{1}{2}\|\gamma(x) + \kappa\|^2_{K_I}. \tag{6.27}$$

The mapping $v \mapsto y_{\mathrm{IP}}$ is passive with storage function $U_{\mathrm{IP}}(x)$.

6.3.3 Lyapunov Stabilization with the New PID-PBC

In this section, complete the design of the new PID-PBC. The key step to propose the derivative term is to differentiate y_{IP}, defined in (6.26), and group the terms depending on u to – via an inversion of this factor – compute the actual control action. This operation is similar to the one required to ensure well posedness of the closed-loop carried out in Section IV.A of (Zhang et al., 2018) and leads to the following:

Assumption 6.2: The matrix $\mathsf{K}(x) \in \mathbb{R}^{m\times m}$, defined as[3]

$$\mathsf{K}(x) := I_m + K_D(\nabla y_{\mathrm{IP}})^\top g(x), \tag{6.28}$$

with y_{IP} given in (6.26), has *full rank*.

Proposition 6.8: *Fix an assignable equilibrium $x^\star \in \mathcal{E}$. Consider the system (5.1) satisfying Assumptions 6.1 and 6.2. Define the new desired energy function:*

$$H_{\mathrm{dN}}(x) := U_{\mathrm{IP}}(x) + \frac{1}{2}\|y_{\mathrm{IP}}\|^2_{K_D} \tag{6.29}$$

with y_{IP} and $U_{\mathrm{IP}}(x)$ given in (6.26) and (6.27), respectively. Let the control law be given by the PID-PBC

$$u = -\mathsf{K}^{-1}(x)\left\{K_P y_{\mathrm{IP}} + K_I[\gamma(x) + \kappa] + K_D(\nabla y_{\mathrm{IP}})^\top F(x)\nabla H(x)\right\}. \tag{6.30}$$

---

2 The subscript $(\cdot)_{\mathrm{IP}}$ is used to underscore that these mappings are the projections of the corresponding $(\cdot)_{\mathrm{I}}$ mappings onto the manifold $\mathcal{M}_\kappa$.
3 To simplify the notation, and with some abuse of notation, we use the symbol ∇y_{IP} to denote the gradient of the right hand side of (6.26).

(i) If

$$(\nabla^2 H_{\mathrm{dN}})^\star > 0. \tag{6.31}$$

Then, the closed-loop system (5.1)–(6.30) *has a stable equilibrium at* $x^\star$, *with Lyapunov function* $H_{\mathrm{dN}}(x)$.

(ii) The equilibrium is asymptotically stable if y_{IP} is detectable.

(iii) The stability properties are global if $H_{\mathrm{dN}}(x)$ *is radially unbounded.*

The proof of Proposition 6.8 may be found in Borja et al. (2020).

Remark 6.6: The control law (6.30) may be written as

$$\begin{aligned} \dot{x}_c &= y_{\mathrm{wD}}, \\ u &= -K_P y_{\mathrm{IP}} - K_I x_c - K_D \dot{y}_{\mathrm{IP}}, \end{aligned}$$

with $x_c(0) = \kappa + \gamma(0)$. Hence, it is a PID that relies on the feedback of *two different outputs*, the standard y_{wD} for the integral action and the new output y_{IP} for the derivative and proportional terms.

Remark 6.7: The relation between PID-PBC and other PBC techniques, such as interconnection and damping assignment (IDA) PBC and CbI, can be found in Borja et al., 2020 and Zhang et al. (2018), respectively.

6.4 Examples

The following examples illustrate the applicability of the PID-PBCs to stabilize pH systems.

6.4.1 A Microelectromechanical Optical Switch (Continued)

Consider the microelectromechanical optical switch studied in Section 5.4.2. The set of assignable equilibria for this system is

$$\mathcal{E} = \left\{ x \in \mathbb{R}^3 \mid x_2 = 0, x_3 = (c_0 + x_1)\sqrt{2ac_1 x_1} \right\},$$

where the control objective is to stabilize the actuator at a desired position $x_1^\star > 0$. To this end, we consider

$$y_{\mathrm{wD}} = \dot{x}_3, \qquad \gamma(x) = x_3.$$

Furthermore, for this example,

$$\kappa = -x_3^\star - \left(\frac{\partial H}{\partial x_3} \right)^\star. \tag{6.32}$$

Fix $K_D = 0$. Accordingly, the PID-PBC (6.7) takes the form

$$u = -K_P\dot{x}_3 - K_I\left(x_3 - x_3^\star\right) + \left(\frac{\partial H}{\partial x_3}\right)^\star,$$

which can be rewritten as

$$u = -\left(1 + \frac{K_P}{r}\right)^{-1}\left\{\frac{K_P}{r}\frac{\partial H}{\partial x_3} + K_I\left(x_3 - x_3^\star\right) - \left(\frac{\partial H}{\partial x_3}\right)^\star\right\}.$$

Some simple computations show that

$$\left(\nabla H_d\right)^\star = 0, \qquad (\nabla^2 H_d)^\star > 0,\ \forall K_I > 0.$$

Therefore, $x^\star$ is an isolated minimum of $H_{\mathrm{I}}(x)$, and a stable equilibrium for the closed-loop system.

To prove asymptotic stability, note that

$$y_{\mathrm{wD}} = 0 \iff \dot{x}_3 = 0 \iff \frac{\partial H}{\partial x_3} - \left(\frac{\partial H}{\partial x_3}\right)^\star + K_I\left(x_3 - x_3^\star\right) = 0. \tag{6.33}$$

Hence, differentiating the latter expression yields

$$\left(\frac{\partial^2 H}{\partial x_3 x_1}\right)\dot{x}_1 = 0,$$

which in combination with (6.33) implies that

$$\dot{x}_1 = 0 \iff x_2 = 0 \implies \dot{x}_2 = 0 \iff \frac{\partial H}{\partial x_1} = 0. \tag{6.34}$$

Then, substituting (6.34) in (6.33), we conclude that the passive output is detectable, which implies that $x^\star$ is an asymptotically stable equilibrium for the closed-loop system.

6.4.2 Boost Converter

Consider the simplified model of a boost converter, given by (5.1) with

$$F = -\begin{bmatrix} r_1 & 0 \\ 0 & r_2 \end{bmatrix},\quad g(x) = \begin{bmatrix} -x_2 \\ x_1 \end{bmatrix},\quad H(x) = \frac{1}{2}|x|^2 - \frac{V_s}{r_1}x_1,$$

where r_1, r_2, and V_s are positive constants, and the state variables are subject to the physical constraint $x_i > 0$, for $i = 1, 2$.

The set of assignable equilibria for this system is characterized by

$$\mathcal{E} = \left\{x \in \mathbb{R}^2 \mid x_1\left(V_s - r_1 x_1\right) - r_2 x_2^2 = 0\right\},$$

and

$$u^\star = r_2 \frac{x_2^\star}{x_1^\star}.$$

The control objective is to stabilize the system at a desired voltage $x_2^\star > 0$.

Note that

$$y_{\mathrm{wD}} = -g^\top(x) F \dot{x} = -\frac{1}{r_1} x_2 \dot{x}_1 + \frac{1}{r_2} x_1 \dot{x}_2,$$

which is not *integrable*. A solution to this problem is given by the use of multipliers, see Section 6.2.3. In particular, for this example, the conditions **C1** and **C2** of Proposition 6.5 are verified by

$$\Lambda(x) = \begin{bmatrix} r_1 x_1 \\ r_2 x_2 \end{bmatrix}.$$

Then, since $n - m = 1$, (6.23) holds. The latter guarantees the existence of $M(x)$ such that the passive output $\overline{y} = M(x) y_{\mathrm{wD}}$ is *integrable*. In particular,

$$M(x) = \frac{1}{x_1 x_2}$$

yields

$$\overline{y} = -\frac{1}{r_1 x_1} \dot{x}_1 + \frac{1}{r_2 x_2} \dot{x}_2, \qquad \gamma(x) = -\frac{1}{r_1} \ln(x_1) + \frac{1}{r_2} \ln(x_2).$$

Consider $K_D = 0$. Hence, the PI-PBC (6.7) takes the form

$$\begin{aligned} \bar{u} = -x_1 x_2 \Bigg\{ & K_I \left[-\frac{1}{r_1} \left(\ln(x_1) - \ln(x_1^\star) \right) + \frac{1}{r_2} \left(\ln(x_2) - \ln(x_2^\star) \right) \right] \\ & + K_P \left(-\frac{1}{r_1 x_1} \dot{x}_1 + \frac{1}{r_2 x_2} \dot{x}_2 \right) - u^\star \Bigg\}, \end{aligned}$$

which is equivalent to

$$\begin{aligned} \bar{u} = -x_1 x_2 \left[1 + K_P \left(\frac{1}{r_1 x_1^2} + \frac{1}{r_2 x_2^2} \right) \right]^{-1} & \left\{ K_P \left(\frac{1}{x_1} \frac{\partial H}{\partial x_1} + \frac{1}{x_2} \frac{\partial H}{\partial x_1} \right) - u^\star \right. \\ & \left. + K_I \left[-\frac{1}{r_1} \left(\ln(x_1) - \ln(x_1^\star) \right) + \frac{1}{r_2} \left(\ln(x_2) - \ln(x_2^\star) \right) \right] \right\}. \end{aligned}$$

Moreover, $(\nabla H_d)^\star = 0$ and – depending on the parameters of the system – there exists K_I large enough such that $(\nabla^2 H_d)^\star > 0$. Therefore, $x^\star$ is an isolated minimum of $H_d(x)$, and a stable equilibrium for the closed-loop system.

6.4.3 Two-Dimensional Controllable LTI Systems

Consider a two-dimensional LTI system in the controllable canonical form, i.e.

$$\dot{x} = \begin{bmatrix} 0 & 1 \\ a_1 & a_2 \end{bmatrix} \begin{bmatrix} x_1 \\ x_2 \end{bmatrix} + \begin{bmatrix} 0 \\ 1 \end{bmatrix} u, \tag{6.35}$$

where the control objective consists in stabilizing the system at the origin. Toward this end, we express (6.35) as a pH system with[4]

$$F = \begin{bmatrix} 0 & -\text{sign}(a_2) \\ \text{sign}(a_2) & -|a_2| \end{bmatrix}, \quad g = \begin{bmatrix} 0 \\ 1 \end{bmatrix}, \quad H(x) = \frac{1}{2} x^\top Q x, \tag{6.36}$$

where[5] $Q = \text{diag}\{\text{sign}(a_2)a_1, -\text{sign}(a_2)\}$. Consider

$$y_{\text{wD}} = -\text{sign}(a_2)x_2, \qquad \gamma(x) = -\text{sign}(a_2)x_1.$$

Hence, some straightforward computations show that, for this example, $\kappa = 0$, and – considering $y_{\text{IP}} = y_{\text{wD}}$ – Assumption 6.2 holds for $K_D \neq 1$. Accordingly, the PID-PBC (6.7) takes the form

$$u = \frac{\text{sign}(a_2)}{1 - \text{sign}(a_2)K_D} \left\{ (K_I + a_1 K_D)x_1 + (K_P + a_2 K_D)x_2 \right\}. \tag{6.37}$$

Therefore, the system (6.35) in closed-loop with (6.37) reduces to

$$\dot{x} = A_{\text{cl}}x, \qquad A_{\text{cl}} := \begin{bmatrix} 0 & 1 \\ \dfrac{a_1 + \text{sign}(a_2)K_I}{1 - \text{sign}(a_2)K_D} & \dfrac{a_2 + \text{sign}(a_2)K_P}{1 - \text{sign}(a_2)K_D} \end{bmatrix},$$

where the real part of the eigenvalues of A_{cl} is negative for nonnegative gains K_P, K_I, K_D satisfying

$$\begin{aligned} K_D > 1, (a_1 + K_I) \geq 0, \quad &\text{if} \quad a_2 \geq 0, \\ (a_1 - K_I) < 0, \quad &\text{if} \quad a_2 < 0. \end{aligned} \tag{6.38}$$

Remark 6.8: The controllability condition of an LTI is not enough to ensure that it can be stabilized via PI-PBC, as it has been discussed in Borja et al. (2016). In fact, in the example above, it follows from (6.38) that the derivative action is necessary to stabilize the system when $a_2 \geq 0$.

---

4 Where we consider $\text{sign}(0) = 1$.

5 Note that the lower boundedness of $H(x)$ depends on the signs of the parameters a_1, a_2.

6.4.4 Control by Interconnection vs. PI-PBC

Consider a pH system, with

$$F = \begin{bmatrix} 0 & 1 \\ -1 & 0 \end{bmatrix}, \quad g(x) = \begin{bmatrix} x_1 \\ 0 \end{bmatrix}, \quad H(x) = \tfrac{1}{2}(x_1 + x_2)^2, \tag{6.39}$$

where the objective is to stabilize the origin. Note that

$$y_{\mathrm{wD}} = -g^\top(x)F^{-\top}\dot{x} = -g^\top(x)\nabla H(x) = x_1\dot{x}_2, \tag{6.40}$$

which is not *integrable*. To overcome this issue, consider the selection of the parameters $w(x)$ and $D(x)$ provided in Proposition 6.4 with

$$\phi(x) = 0, \;\; \alpha(x) = \begin{bmatrix} -x_1 \\ x_1 \end{bmatrix}, \quad \rho(x) = x_1.$$

Then,

$$y_{\mathrm{wD}} = \beta^\top(x)\dot{x} = x_1\dot{x}_1, \;\; \beta(x) := \begin{bmatrix} x_1 \\ 0 \end{bmatrix}.$$

Hence,

$$\gamma(x) = \frac{1}{2}x_1^2$$

verifies $\dot{\gamma} = y_{\mathrm{wD}}$. Moreover, in this example $\kappa = 0$. Fix $K_D = 0$. Hence, the PI-PBC (6.7) takes the form

$$u = -K_P x_1\dot{x}_1 - \frac{1}{2}K_I x_1^2 + u^\star, \tag{6.41}$$

where $u^\star$ is an arbitrary negative constant. The control law (6.41) can be rewritten exclusively in terms of x as

$$u = -\left(1 + K_P x_1^2\right)^{-1}\left[K_P x_1(x_1 + x_2) + \frac{1}{2}K_I x_1^2 - u^\star\right].$$

Furthermore, some straightforward computations show that

$$(\nabla H_d)^\star = 0, \qquad (\nabla^2 H_d)^\star > 0, \forall u^\star < 0.$$

Accordingly, the origin is an isolated minimum of $H_d(x)$ and a stable equilibrium of the closed-loop system. To prove that the equilibrium is asymptotically stable, notice that

$$y_{\mathrm{wD}} = 0 \iff x_1\dot{x}_1 = 0,$$

which implies that either x_1 or $\dot{x}_1$ is zero. If $x_1 = 0$, then $\dot{x}_1 = 0$ implies $x_2 = 0$. On the other hand, if $\dot{x}_1 = 0$, but $x_1 \neq 0$, then

$$x_1 + x_2 + x_1 u^\star - \frac{1}{2}K_I x_1^3 = 0. \tag{6.42}$$

Differentiating the equation above, we get

$$\dot{x}_2 = 0 \iff x_1 = -x_2. \tag{6.43}$$

Thus, replacing (6.43) in (6.42) yields

$$x_1\left(u^\star - \frac{1}{2}K_I x_1^2\right) = 0. \tag{6.44}$$

Since $u^\star < 0$, the only solution to the equation above is $x_1 = 0$. Accordingly, y_{wD} is detectable and the origin is an asymptotically stable equilibrium for the closed-loop system.

6.4.5 The Use of the Derivative Action

Consider a pH system, with

$$F = \begin{bmatrix} 0 & 1 & 0 \\ -1 & 0 & -1 \\ 0 & 1 & -1 \end{bmatrix}, \quad g = \begin{bmatrix} 0 & 0 \\ 1 & 0 \\ 0 & 1 \end{bmatrix},$$

$$H(x) = a_1(\cos x_1 + 1) - \tfrac{1}{2}a_2x_2^2 + a_3x_3,$$

where a_1, a_2, a_3 are positive constants, and the control objective is to stabilize the point $x^\star = 0$. Note that $\mathcal{R}^\star(\nabla H)^\star \neq 0$, which implies that this system is affected by the dissipation obstacle. Thus, to stabilize this system via a PID-PBC, it is necessary to consider a passive output with relative degree zero.

Consider[6]

$$y_{\text{wD}} = -g^\top F^\top \dot{x} = \begin{bmatrix} \dot{x}_1 \\ \dot{x}_3 - \dot{x}_1 \end{bmatrix}, \quad \gamma(x) = \begin{bmatrix} x_1 \\ x_3 - x_1 \end{bmatrix}, \quad K_I = \text{diag}\left\{k_{i1}, k_{i2}\right\} > 0.$$

Thus,

$$\kappa = \begin{bmatrix} \kappa_1 \\ \kappa_2 \end{bmatrix} = \begin{bmatrix} -\frac{a_3}{k_{i1}} \\ -\frac{a_3}{k_{i2}} \end{bmatrix}.$$

Therefore,

$$(\nabla^2 U_{\text{IP}})^\star = \begin{bmatrix} k_{i1} - a_1\cos x_1 & 0 & 0 \\ 0 & -a_2 & 0 \\ 0 & 0 & 0 \end{bmatrix} + k_{i2}\begin{bmatrix} 1 & 0 & -1 \\ 0 & 0 & 0 \\ -1 & 0 & 1 \end{bmatrix}.$$

Note that, since $a_2 > 0$, $(\nabla^2 U_{\text{IP}})^\star$ has a negative eigenvalue no matter the values of k_{i1} and k_{i2}.

6 For simplicity, we choose K_I diagonal. However, the fact that this system cannot be stabilized by the control law (6.7), with $K_D = 0$, is independent of the selection of K_I.

The analysis above leads to the conclusion that the PID-PBC (6.7), with $K_D = 0$, cannot render stable the desired equilibrium. To overcome this issue, consider

$$y_{\mathrm{IP}} = \begin{bmatrix} -a_2 x_2 \\ a_3 + k_{i2}(x_3 - x_1 + \kappa_2) \end{bmatrix}, \qquad K_D = \mathrm{diag}\left\{k_{d1}, k_{d2}\right\},$$

where $k_{d2} > 0$ and $a_2 k_{d1} > 1$. Therefore, Assumption 6.2 is satisfied. Moreover,

$$\left(\nabla U_{\mathrm{IP}}\right)^{\star} = 0 \implies \left(\nabla H_{\mathrm{dN}}\right)^{\star} = 0,$$

and $\left(\nabla^2 H_{\mathrm{dN}}\right)^{\star} > 0$ for $k_{i1} > a_1$. Accordingly, there exist positive gain matrices K_I and K_D such that $x^{\star}$ is an isolated minimum of $H_{\mathrm{dN}}(x)$. Additionally, we have the following chain of implications:

$$\begin{aligned} y_{\mathrm{IP}} = 0 \quad &\implies \quad \begin{cases} x_2 = \dot{x}_2 = 0, \\ a_3 + k_{i2}(x_3 - x_1 + \kappa_2) = 0, \end{cases} \\ &\implies \quad x_1 = x_3, \\ &\implies \quad \tfrac{k_{i1}}{a_1} x_1 = \sin x_1. \end{aligned}$$

The latter must hold for any value of k_{i1} and a_1. Hence, $x_1 = 0$ implies $x_3 = 0$. Consequently, y_{IP} is detectable, and the PID-PBC (6.7) – with K_I and K_D satisfying the conditions mentioned above – renders $x^{\star}$ asymptotically stable.

Bibliography

P. Borja, R. Cisneros, and R. Ortega. A constructive procedure for energy shaping of port-Hamiltonian systems. *Automatica*, 72: 230–234, 2016.

P. Borja, R. Ortega, and J. M. A. Scherpen. New results on stabilization of port-Hamiltonian systems via PID passivity-based controllers. *IEEE Transactions on Automatic Control*, 66(2): 625–636 2020.

F. Castaños, B. Jayawardhana, R. Ortega, and E. García-Canseco. Proportional plus integral control for set point regulation of a class of nonlinear RLC circuits. *Circuits, Systems and Signal Processing*, 28(4): 609–623, 2009.

C. A. Desoer and M. Vidyasagar. *Feedback Systems: Input-Output Properties*. Academic Press, New York, 2009.

C. M. Kellett and P. M. Dower. Input-to-state stability, integral input-to-state stability, and $\mathcal{L}_2$-gain properties: qualitative equivalences and interconnected systems. *IEEE Transactions on Automatic Control*, 61(1): 3–17, 2016.

H. Khalil. *Nonlinear Systems*. Prentice-Hall, Upper Saddle River, NJ, 2002.

B. M. Maschke, R. Ortega, and A. J. van der Schaft. Energy-based Lyapunov functions for forced Hamiltonian systems with dissipation. *IEEE Transactions on Automatic Control*, 45(8): 1498–1502, 2000.

R. Ortega and E. García-Canseco. Interconnection and damping assignment passivity-based control: a survey. *European Journal of Control*, 10(5): 432–450, 2004.

R. Ortega. Comments on recent claims about trajectories of control systems valid for particular initial conditions. *Asian Journal of Control*, 1–8, 2021.

M. Spivak. *Comprehensive Introduction to Differential Geometry*. Perish, Inc., 3rd edition, 1999.

A. J. van der Schaft. *L_2-Gain and Passivity Techniques in Nonlinear Control*. Springer-Verlag, Berlin, 3rd edition, 2016a.

A. J. van der Schaft. Interconnections of input-output Hamiltonian systems with dissipation. In *IEEE Conference on Decision and Control*, pages 4686–4691, Las Vegas NV, USA, 2016b.

M. Vidyasagar. *Nonlinear Systems Analysis*. 2nd edition. Prentice-Hall, Englewood Cliffs, NJ, 1993.

M. Zhang, P. Borja, R. Ortega, Z. Liu, and H. Su. PID passivity-based control of port-Hamiltonian systems. *IEEE Transactions on Automatic Control*, 63(4): 1032–1044, 2018.

7

Underactuated Mechanical Systems

In this chapter, we apply the PID-PBC methodology to control (simple) underactuated mechanical systems – that is, mechanical systems where the number of controls is strictly smaller than the number of degrees of freedom (DOF). Similarly to the developments of Chapter 6, the objective is to stabilize an equilibrium point wrapping up a PID controller around a passive output. The stability analysis is carried out, also, via the construction of a Lyapunov function, but with the significant differences explained in Section 7.2.

Further details of the material reported in this chapter may be found in Donaire et al. (2016a), Romero et al. (2016, 2018), Gandhi et al. (2016), and Ortega et al. (2017).

7.1 Historical Review and Chapter Contents

Interestingly, it is with the study of mechanical systems that the idea of PBC – as a technique to shape the energy of the system and add damping – first emerged in Ortega et al. (2002). Analysis of the stability of mechanical systems by studying their energy function is a well-established technique whose roots date back to the work of Lagrange, Dirichlet, and Lord Kelvin – see Koditschek (1989) for a fascinating review of this circle of ideas. In this section, we briefly discuss the history of PID-PBC and outline the contents of the chapter.

7.1.1 Potential Energy Shaping of Fully Actuated Systems

In the control context, energy shaping was first used by Takegaki and Arimoto in the seminal paper (Takegaki and Arimoto, 1981) who proposed

PID Passivity-Based Control of Nonlinear Systems with Applications, First Edition.
Romeo Ortega, José Guadalupe Romero, Pablo Borja, and Alejandro Donaire.

Published 2021 by John Wiley & Sons, Inc.

to shape the potential energy and to add damping to solve the point-to-point positioning task for a *fully actuated* robot manipulator. Let us illustrate this with the simple pendulum example, whose total energy is given by

$$H(q, \dot{q}) = \underbrace{\frac{1}{2} m\ell^2 \dot{q}^2}_{T(\dot{q})} + \underbrace{mg\ell[1 - \cos(q)]}_{V(q)},$$

with $q(t) \in \mathbb{R}$, $T : \mathbb{R} \to \mathbb{R}_{\geq 0}$, $V : \mathbb{R} \to \mathbb{R}$ the kinetic and potential energies, respectively, and m, g, and ℓ are positive constants. The dynamics of the pendulum is of the form

$$m\ell^2 \ddot{q} + V'(q) = \tau,$$

with the torque τ the control input. A desired equilibrium $(q, \dot{q}) = (q^\star, 0)$ can be asymptotically stabilized shaping the potential energy and adding some damping via

$$\tau = V'(q) - V_d'(q) - k_p \dot{q}, \tag{7.1}$$

with $V_d(q)$ verifying

$$q^\star = \arg\min V_d(q), \tag{7.2}$$

and $k_p > 0$. A suitable Lyapunov function is the new total energy $H_d(q, \dot{q}) = T(\dot{q}) + V_d(q)$. A particularly interesting choice of desired potential energy is

$$V_d(q) = V(q) + \frac{k_i}{2}\left[q - q^\star + \frac{1}{k_i} V'(q^\star)\right]^2,$$

which satisfies (7.2) for $k_i > 0$ sufficiently large. The resulting control law is

$$\tau = -k_p \dot{q} - k_i \tilde{q} + V'(q^\star),$$

that is, a simple proportional-derivative (PD) law around the position q. This result can be directly extended to n-DOF mechanical systems.[1]

Not surprisingly, though unknown to the previous authors, the key property underlying the success of such a simple scheme is the *passivity* of the system dynamics. Indeed, it is easy to see that $\dot{H} = \dot{q}\tau$, establishing the passivity of the map $\tau \mapsto \dot{q}$. With respect to this output, the feedback law (7.1) *is a PID-PBC* plus a constant term. As proved in Proposition 2.2.5 of (Ortega et al., 1998) a broad class of Euler–Lagrange (EL) systems define passive maps from the external forces to the derivative of the generalized coordinates, which in the case of mechanical systems are the coordinate velocities.

1 Interestingly Jonckheere (1981), independently of Takegaki and Arimoto, suggested also the use of PD-like energy shaping and damping injection controllers for stabilization of a class of EL systems, which includes mechanical, electrical, and electromechanical systems.

7.1.2 Total Energy Shaping of Underactuated Systems

While fully actuated mechanical systems admit an arbitrary shaping of the potential energy by means of feedback, and therefore stabilization to any desired equilibrium, this is in general not possible for underactuated systems. In certain cases, this problem can be overcome by also modifying the kinetic energy of the system. This idea of *total energy shaping* was proposed in Ailon and Ortega (1993) where the first solution to the problem of position feedback stabilization of robots with flexible joints was solved modifying both the kinetic and potential energies of the manipulator and adding damping through the controller. See Section 2.4 for an interpretation of this scheme as CbI.

It is also possible to modify the total energy and add damping via *static* state-feedback, which is the approach adopted in the method of controlled Lagrangians (CL) (Bloch et al., 2000, 2001) and IDA-PBC (Ortega et al., 2002), see also Blankenstein et al. (2002) where it is shown that the class of mechanical pH systems considered in IDA-PBC strictly contains the EL systems proposed in the CL method, and the closely related work (Fujimoto and Sugie, 2001). In both cases, stabilization (of a desired equilibrium) is achieved identifying the class of systems – EL for CL or pH for IDA-PBC – that can possibly be obtained via feedback. The conditions under which such a feedback law exists are called *matching conditions*, and consist of a set of nonlinear PDEs. In case these PDEs can be solved the original control system and the target dynamic system are said to *match* and the solutions of the PDEs identify the assignable Hamiltonian or Lagrangian functions, respectively.

7.1.3 Two Formulations of PID-PBC

Similarly to the IDA and CL total energy shaping methods, the design of PID-PBCs may proceed from a pH or an EL description of the system.[2] The first reported result on this topic (Donaire et al., 2016a) was done using the EL formulation. To achieve the control objective, it was necessary to introduce in the design a first partial linearization (Spong, 1994) feedback. As is well known, the exact cancellation of nonlinearities, intrinsic to feedback linearization, is nonrobust. To overcome this obstacle, in Romero et al. (2016), the PID-PBC design problem was formulated and solved, in the pH framework. It turned out that, the partial feedback linearization could be avoided introducing a suitable change of coordinates in the momenta. This

2 We recall that for EL systems the state is $(q, \dot{q})$, while for pH ones it is (q, p), with p the momenta defined as $p = M(q)\dot{q}$. See Appendix D for further details.

change of coordinates was used before for speed observation and output feedback stabilization of mechanical systems in Venkatraman et al. (2010). Working out in this new system representation, the task of identifying the passive outputs was trivialized.

However, the mathematical manipulations needed to establish the result were quite technical with an unclear interpretation. This difficulty was overcome in Romero et al. (2018), where the pH-based results of (Romero et al., 2016), were reinterpreted in the EL context. It was shown that working with velocities, instead of momenta, the passive outputs, and their associated storage functions, had a clear interpretation. It should be mentioned, however, that as shown in Romero et al. (2016), carrying out the stability analysis in the EL framework is far more complicated than the one done for a pH representation.

In Section 7.2, we first explain, in words, the procedure followed to design the PID-PBC, highlighting the similarities and differences with respect to the material presented in Chapter 6. Then, in Section 7.3, we present the main results proceeding from the pH formulation. In Section 7.4, we discuss the previous results in an EL formulation and present some additional extensions of the results in Section 7.5. Sections 7.6 and 7.7 are devoted to give some practical examples and the extension of the method to constrained EL systems, respectively.

7.2 Shaping the Energy with a PID

In this section, we see how the PID-PBC methodology allows us to identify a class of mechanical systems for which it is possible to shape the total energy of the system without solving PDEs. The main difference with respect to the IDA-PBC and the CL methods is that *we do not aim* at the preservation in closed-loop of the mechanical structure of the system – the condition that gives rise to the matching PDEs. Instead, we concentrate our attention on the energy shaping objective only and give conditions on the mechanical system such that this is achievable without solving PDEs.

The approach proposed in this chapter differs also from the one pursued in Chapter 6, which we recall proceeds as follows:

S1 Identify all outputs which are passive with storage function and the Hamiltonian of the pH system $H(x)$ – see Proposition 5.1 – and wrap the PID around one of the passive outputs.

S2 Construct the function $U(x, x_c) := H(x) + H_c(x_c)$, which in view of the passivity properties of system and controller is nonincreasing.

S3 This function is, in general, not positive definite (with respect to the desired equilibrium). To ensure the latter property, we project the state of the PID into the state of the system via the generation of a first integral relating the controller and the system's state of the form $x_c = \gamma(x)$, see (6.5). That is, to ensure that the foliation

$$\mathcal{M}_\kappa := \{(x, x_c) \in \mathbb{R}^n \times \mathbb{R}^m \mid x_c = \gamma(x) + \kappa\}, \tag{7.3}$$

with $\kappa \in \mathbb{R}$, is *invariant*. To accomplish the latter task, it is necessary to solve the PDE (6.6), which is usually far from trivial.

S4 Among all solutions $\gamma(x)$ of the PDE select one that assigns the desired minimum to the function $H_d(x) := U(x, \gamma(x) + \kappa)$, which will then qualify as a *bona fide* Lyapunov function for the closed-loop, see (6.3) and Proposition 6.2.

In the approach followed in this chapter, we first construct a passive output $y_d = h_d(q, p) \in \mathbb{R}^m$ – with its corresponding storage function $H_a(q, p)$ – around which we wrap up the PID. Notice that, in contrast to Step **S1** above, the storage function *is not* the system's Hamiltonian. Similar to Step **S2** above, the function $U(q, p, x_c) := H_a(q, p) + H_c(x_c)$ is nonincreasing. To be able to construct a Lyapunov function, similarly to Step **S3** above, we find a function $\gamma : \mathbb{R}^n \to \mathbb{R}^m$ such that

$$\frac{d}{dt}\gamma(q) = h_d(q, p) = \dot{x}_c. \tag{7.4}$$

We underscore the fact that we have selected $\gamma(q)$ as a function only of q. This is because, as we show below, this is the only coordinate that needs to be "shaped."

The solution of the PDE associated with (7.4) is *avoided* imposing some restrictions on the structure of the mechanical system – including some integrability conditions on some terms of the inertia matrix – that reduce the solution of the PDE to a simple integration. To complete the design, we exploit the fact that mechanical systems have a pH structure with a particular Hamiltonian function, i.e. the total energy, which is of the form

$$H(q, p) := \frac{1}{2}\, p^\top\, M^{-1}(q)\, p + V(q), \tag{7.5}$$

where $q(t) \in \mathbb{R}^n$, $p(t) \in \mathbb{R}^n$ are the generalized position and momenta, respectively, $M : \mathbb{R}^n \to \mathbb{R}^{n \times n}$, is the positive definite inertia matrix and $V : \mathbb{R}^n \to \mathbb{R}$ is the potential energy function – see Appendix D.2. Motivated by this fact, we *fix the structure* of the Lyapunov function candidate for the closed-loop to be of the form

$$H_d(q, p) := \frac{1}{2}\, p^\top\, M_d^{-1}(q)\, p + V_d(q), \tag{7.6}$$

where $M_d : \mathbb{R}^n \to \mathbb{R}^{n\times n}$ and $V_d : \mathbb{R}^n \to \mathbb{R}$. That is, we impose the constraint on $\gamma(q)$ to satisfy

$$U(q, p, \gamma(q)) = H_d(q, p).$$

If, among all solutions $\gamma(q)$ of (7.4), we can find one that ensures that $M_d(q)$ is *positive definite* and $V_d(q)$ satisfies (7.2), then $H_d(q, p)$ is a *bona fide* Lyapunov function for the desired equilibrium $(q, p) = (q^\star, 0)$ of the closed-loop system.

Although the procedure described above seems a bit contrived, it has been possible to solve many interesting benchmark examples with it – including, among others, the inertia wheel (Ortega et al., 2002), the inverted pendulum on a cart (Bedrossian and Spong, 1995), the 2-DOF spider crane, the 4-DOF overhead crane (Donaire et al., 2016), and the spherical pendulum on a puck (Shiriaev et al., 2014). Some geometric interpretations of the conditions imposed on the system, in particular, its connection with the matching conditions of the CL method, are reported in Mehra et al. (2017).

7.3 PID-PBC of Port-Hamiltonian Systems

The pH representation of mechanical systems is given in Section D.2 of Appendix D which, for ease of reference, we repeat below

$$\begin{bmatrix} \dot{q} \\ \dot{p} \end{bmatrix} = \begin{bmatrix} 0 & I_n \\ -I_n & 0 \end{bmatrix} \nabla H(q, p) + \begin{bmatrix} 0 \\ G(q) \end{bmatrix} \tau, \tag{7.7}$$

with total energy function $H : \mathbb{R}^n \times \mathbb{R}^n \to \mathbb{R}$ given by (7.5), input $\tau \in \mathbb{R}^m$, and $G : \mathbb{R}^n \to \mathbb{R}^{n\times m}$ the full-rank input matrix.

Some simple calculations show that $\dot{H} = \tau^\top G^\top(q)\dot{q}$ the mapping $\tau \mapsto G^\top(q)\dot{q}$ is cyclo-passive.

7.3.1 Assumptions on the System

In this section, we present some assumptions on the pH system (7.7) necessary to identify the passive outputs used in the PID-PBC. To simplify the derivations, we make the assumption that the input matrix $G(q)$ is of the form[3]

$$G = \begin{bmatrix} 0 \\ I_m \end{bmatrix}, \tag{7.8}$$

3 It is possible to show that there exists a change of state and input coordinates such that this assumption holds *if and only if* the distribution spanned by the columns of the matrix $G(q)$ is *involutive* – see Nijmeijer and van der Schaft (1990).

where $s := n - m$. Conformally, we partition the inertia matrix as

$$M(q) = \begin{bmatrix} m_{uu}(q) & m_{au}^{\top}(q) \\ m_{au}(q) & m_{aa}(q) \end{bmatrix}, \tag{7.9}$$

where $m_{uu} : \mathbb{R}^n \to \mathbb{R}^{s\times s}, m_{aa} : \mathbb{R}^n \to \mathbb{R}^{m\times m}$ and $m_{au} : \mathbb{R}^n \to \mathbb{R}^{s\times m}$, and also partition the generalized coordinates as $q = \text{col}(q_u, q_a)$, with $q_u \in \mathbb{R}^s$ and $q_a \in \mathbb{R}^m$.

The following *structural* assumption is made on the system.

Assumption 7.1: Consider the mechanical system (7.7):

(i) The inertia matrix depends only on the unactuated variables q_u.
(ii) The $(2, 2)$ subblock of the inertia matrix is constant.
(iii) The potential energy can be written as

$$V(q) = V_a(q_a) + V_u(q_u).$$

7.3.2 A Suitable Change of Coordinates

The first step in the design of the PID-PBC is to introduce the change of coordinates $(q, \mathbf{p}) \mapsto (q, T^{\top}(q)p)$, where $T : \mathbb{R}^n \to \mathbb{R}^{n\times n}$ is a *full rank* factorization of the inverse inertia matrix, that is,

$$M^{-1}(q) = T(q)T^{\top}(q). \tag{7.10}$$

As shown in Venkatraman et al. (2010), it transforms (7.7) into

$$\begin{bmatrix} \dot{q} \\ \dot{\mathbf{p}} \end{bmatrix} = \begin{bmatrix} 0 & T(q) \\ -T^{\top}(q) & J(q, \mathbf{p}) \end{bmatrix} \nabla W(q, \mathbf{p}) + \begin{bmatrix} 0 \\ T^{\top}(q)G \end{bmatrix} \tau, \tag{7.11}$$

with Hamiltonian $W : \mathbb{R}^n \times \mathbb{R}^n \to \mathbb{R}$

$$W(q, \mathbf{p}) = \frac{1}{2}|\mathbf{p}|^2 + V(q),$$

and the jk-th element of the *skew-symmetric* matrix $J : \mathbb{R}^n \times \mathbb{R}^n \to \mathbb{R}^{n\times n}$ given by

$$(J(q, p))_{jk} = -p^{\top}[(T(q))_j, (T(q))_k], \ j, k \in \bar{n}. \tag{7.12}$$

The key observation is that (i) and (ii) of Assumption 7.1 ensure the existence of a factorization (7.10) of the form

$$T(q_u) = \begin{bmatrix} T_1(q_u) & 0 \\ T_2(q_u) & T_3 \end{bmatrix}, \tag{7.13}$$

where $T_1 : \mathbb{R}^s \to \mathbb{R}^{s\times s}$, $T_2 : \mathbb{R}^s \to \mathbb{R}^{m\times s}$, and $T_3 \in \mathbb{R}^{m\times m}$ is *constant*.

An important corollary of (7.10) and (7.13) is that the matrix $J(q,p)$ of the transformed system (7.11) takes the form

$$J(q_u,p) = \begin{bmatrix} J_a(q_u,p) & 0 \\ 0 & 0 \end{bmatrix}, \tag{7.14}$$

for some $J_a : \mathbb{R}^s \times \mathbb{R}^n \to \mathbb{R}^{s \times s}$. This fact is easily established from (7.12) noting that for a $T(q_u)$ of the form (7.13) the Lie brackets take the form

$$[(T(q))_i, (T(q))_j] = \left[(\nabla T(q_u))_j \;\; 0\right](T(q_u))_i - \left[(\nabla T(q_u))_i \;\; 0\right](T(q_u))_j,$$

and doing some simple calculations.

The following Lemma will prove instrumental for generating the passive outputs needed for the design of the PID-PBC. Its proof is established with some simple calculations proceeding from the identity $M(q)T(q)T^\top(q) = I_n$.

Lemma 7.1: *Consider the factorization* (7.10) *with* $M(q)$ *and* $T(q)$ *partitioned as* (7.9) *and* (7.13), *respectively. The following identities hold*

$$[T_1(q_u)T_1^\top(q_u)]^{-1} = m_{uu}(q_u) - m_{au}^\top(q_u)m_{aa}^{-1}m_{au}(q_u) \tag{7.15a}$$

$$T_2(q_u)T_1^{-1}(q_u) = -m_{aa}^{-1}m_{au}(q_u) \tag{7.15b}$$

$$T_3T_3^\top = m_{aa}^{-1}. \tag{7.15c}$$

Remark 7.1: If the *Riemann symbols* (Eisenhart, 1997) of the inertia matrix are zero then, for *any* factorization $T(q)$, we have

$$[(T(q))_j, (T(q))_k] = 0, \; j,k \in \bar{n}.$$

In view of (7.12), this implies that $J(q,p) = 0$. These systems have been extensively studied in analytical mechanics and have a deep geometric significance (Nijmeijer and van der Schaft, 1990). They belong to the class of systems that are partially linearized via change of coordinates studied in Venkatraman et al. (2010) and, as shown below, fit into the stabilization framework proposed in the section.

7.3.3 Generating New Passive Outputs

The key step of identifying the new passive outputs is given in the following proposition. For notational convenience in the sequel, we partition $\boldsymbol{p} = \mathrm{col}(\boldsymbol{p}_u, \boldsymbol{p}_a)$ with $\boldsymbol{p}_u \in \mathbb{R}^s$ and $\boldsymbol{p}_a \in \mathbb{R}^m$.

Lemma 7.2: *Consider the underactuated mechanical system* (7.11) *with* $M(q)$ *and* $V(q)$ *verifying Assumption 7.1, together with the inner-loop control*

$$\tau = \nabla V_a(q_a) + u. \tag{7.16}$$

Define the output signals

$$y_u := T_2(q_u)\mathbf{p}_u, \quad y_a := T_3\mathbf{p}_a. \tag{7.17}$$

The operators $u \mapsto y_u$ and $u \mapsto y_a$ are passive with storage functions:

$$H_u(q_u, \mathbf{p}_u) = \frac{1}{2}|\mathbf{p}_u|^2 + V_u(q_u), \quad H_a(\mathbf{p}_a) = \frac{1}{2}|\mathbf{p}_a|^2, \tag{7.18}$$

respectively.

Proof. Replacing (7.13), (7.14), and (7.16) in (7.11) yields

$$\begin{aligned}
\dot{q}_u &= T_1(q_u)\mathbf{p}_u \\
\dot{q}_a &= T_2(q_u)\mathbf{p}_u + T_3\mathbf{p}_a \\
\dot{\mathbf{p}}_u &= -T_1^\top(q_u)\nabla V_u(q_u) + J_a(q_u, p)\mathbf{p}_u + T_2^\top(q_u)u \\
\dot{\mathbf{p}}_a &= T_3^\top u.
\end{aligned} \tag{7.19}$$

The time derivatives of the storage functions (7.18) along the solutions of (7.11) verify

$$\dot{H}_a = u^\top y_a \; ; \quad \dot{H}_u = u^\top y_u, \tag{7.20}$$

where we used the fact that $J_a(q_u, p)$ is skew–symmetric. This completes the proof. □

Remark 7.2: It is interesting to note that it is possible to identify the passive outputs y_u and y_a given in (7.17) without appealing to the change of coordinates used above. Indeed, referring to the pH model (7.7), in the original coordinates (q, p), a consequence of (i) and (iii) of Assumption 7.1, and the use of the inner-loop (7.16), is that $\dot{p}_a = u$. Hence, the signal $\breve{y}_a := \Gamma p_a$, with Γ and $n \times n$ positive definite matrix, is a passive output with storage function $\breve{H}_a(p_a) = \frac{1}{2}\|p_a\|^2_\Gamma$. Some lengthy, but straightforward calculations, show that defining $\breve{H}_u(q,p) := H(q,p) - \breve{H}_a(p_a)$ yields $\dot{\breve{H}}_u = u^\top \breve{f}_u(q,p)$, where $\breve{f}_u : \mathbb{R}^n \times \mathbb{R}^n \to \mathbb{R}^m$ is given by

$$\breve{f}_u(q,p) = \begin{bmatrix} 0 & I_m \end{bmatrix} M^{-1}(q)p - \Gamma p_a.$$

Hence, defining $\breve{y}_u := \breve{f}_u(q,p)$, yields another passive output. Moreover, selecting $\Gamma = m_{aa}^{-1}$, and using the change of coordinates $\boldsymbol{p} = T(q)p$, generates *exactly* the passive outputs y_a and y_u and their corresponding storage functions. An interesting open question is whether the availability of the free matrix Γ gives some additional DOF in the design of the PID-PBC.

Remark 7.3: The inner-loop controller (7.16) cancels the term $\nabla V_a(q_a)$, which is a much simpler – and far more robust – operation than the full-fledged partial-feedback linearization used in Proposition 4 of (Donaire et al., 2016a). A particular case, where this cancellation can be avoided is discussed in Section 7.5.2

Remark 7.4: From (7.19), it is clear that, to ensure that the desired unactuated coordinate $q_u^\star$, is part of an equilibrium point $(q, \mathbf{p}) = (q^\star, 0)$, it should satisfy $\nabla V_u(q_u^\star) = 0$.

7.3.4 Projection of the Total Storage Function

To proceed with the controller design, we introduce two constants $k_a \in \mathbb{R}$ and $k_u \in \mathbb{R}$ and define the new output

$$y_d := k_a y_a + k_u y_u, \tag{7.21}$$

around which we add the PID

$$\begin{aligned} \dot{x}_c &= y_d \\ k_e u &= -K_P y_d - K_I x_c - K_D \dot{y}_d, \end{aligned} \tag{7.22}$$

where we notice the presence of a constant $k_e \in \mathbb{R}$ premultiplying u, which will be used for energy-shaping. We also observe that, since the output y_d is relative degree one, it is possible to compute its derivative without differentiation – but imposing a well-posedness condition, articulated below, that ensures the implementation (without singularities) of the control (7.22). See Section 2.2 in Chapter 2 for a thorough discussion on the issue of well posedness.

Some simple calculations, using (7.20) and (7.22), show that the function:

$$U(q, \mathbf{p}, x_c) := k_e[k_a H_a(\mathbf{p}_a) + k_u H_u(q_u, \mathbf{p}_u)] + \frac{1}{2}\|x_c\|^2_{K_I} + \frac{1}{2}\|y_d\|^2_{K_D}, \tag{7.23}$$

verifies

$$\dot{U} \le -\|y_d\|^2_{K_P}.$$

Similarly to Step **S3** of the Section 6.3 in Chapter 6, we need to relate the coordinate x_c with the state of the system, via the generation of an invariant manifold. That is, we need to prove the existence of a function $\gamma : \mathbb{R}^s \to \mathbb{R}^m$ such that $\dot{\gamma} = y_d$. Toward this end, we impose the following:

Assumption 7.2: The rows of the matrix $m_{au}(q_u)$ are gradient vector fields (see Appendix C.2), that is,

$$\nabla(m_{au}(q_u))^i = [\nabla(m_{au}(q_u))^i]^\top, \ \forall i \in \bar{m}. \tag{7.24}$$

Invoking Poincare's Lemma (Khalil, 2002), we have that Assumption 7.2 is equivalent to the existence of a mapping $V_N : \mathbb{R}^s \mapsto \mathbb{R}^m$ such that

$$\dot{V}_N = -m_{aa}^{-1} m_{au}(q_u)\dot{q}_u, \tag{7.25}$$

where we have introduced the matrix $-m_{aa}^{-1}$ to simplify the notation in the sequel.

Lemma 7.3: *Consider the pH system (7.19) with $M(q)$ verifying Assumption 7.2 and the output signals (7.17). The function*

$$\gamma(q) := k_a q_a + (k_u - k_a)V_N(q_u)$$

verifies $\dot{\gamma} = y_d$, with y_d defined in (7.21).

Proof. The proof is completed with the following calculations:

$$\begin{aligned}
\dot{\gamma} &= k_a \dot{q}_a + (k_u - k_a)\dot{V}_N \\
&= k_a[T_2(q_u)\mathbf{p}_u + T_3 \mathbf{p}_a] + (k_a - k_u)m_{aa}^{-1} m_{au}(q_u)\dot{q}_u \\
&= k_a[T_2(q_u)\mathbf{p}_u + T_3 \mathbf{p}_a] + (k_u - k_a)T_2(q_u)T_1^{-1}(q_u)\dot{q}_u \\
&= k_a(y_u + y_a) + (k_u - k_a)y_u,
\end{aligned}$$

where we have used (7.19), (7.25) in the second identity and (7.15b) in the third one and the fact that $T_2(q_u)T_1^{-1}(q_u)\dot{q}_u = \mathbf{p}_u$ in the last one. □

7.3.5 Main Stability Result

Now, it only remains to impose the assumptions required to ensure, on the one hand, well posedness and, on the other hand, guarantee the assignment of a suitable Lyapunov function.

Assumption 7.3: There exist constants $k_e, k_a, k_u \in \mathbb{R}$, $K_D, K_I \in \mathbb{R}^{m\times m}$, $K_D, K_I \geq 0$ such that the following holds:

(i) (Well posedness) $\det[K(q_u)] \neq 0$, $\forall q_u \in \mathbb{R}^s$, where $K : \mathbb{R}^s \to \mathbb{R}^{m\times m}$ is defined as

$$K(q_u) := k_e I_m + k_a K_D T_3 T_3^\top + k_u K_D T_2(q_u) T_2^\top(q_u). \tag{7.26}$$

(ii) (Energy-shaping) The matrix

$$M_d(q_u) = \begin{bmatrix} A(q_u) & k_a k_u T_2^\top(q_u) K_D T_3 \\ k_a k_u T_3^\top K_D T_2(q_u) & D(q_u) \end{bmatrix}^{-1} \tag{7.27}$$

with

$$A(q_u) := k_u^2 T_2^\top(q_u) K_D T_2(q_u) + k_e k_u I_s, \quad D(q_u) := k_e k_a I_m + k_a^2 T_3^\top K_D T_3$$

is *positive definite* and the function

$$V_d(q) := k_e k_u V_u(q_u) + \frac{1}{2} || k_a q_a + (k_u - k_a) V_N(q_u) ||_{K_I}^2, \tag{7.28}$$

is proper and has an *isolated minimum* at $q^\star$.

We are in a position to present the main result of the section. As in Proposition 6.2, the gist of the proof is to show that the projection of the function $U(q, \mathbf{p}, x_c)$, defined in (7.23), on the foliation

$$\mathcal{M}_\kappa := \{(q, x_c) \in \mathbb{R}^n \times \mathbb{R}^m \mid x_c = \gamma(q) + \kappa\}, \tag{7.29}$$

with $\kappa \in \mathbb{R}$, which we know is invariant, equals the desired quadratic function

$$H_d(q, \mathbf{p}) = \frac{1}{2} \mathbf{p}^\top M_d^{-1}(q_u) \mathbf{p} + V_d(q). \tag{7.30}$$

That is, we have to ensure that

$$U(q, \mathbf{p}, \gamma(q) + \kappa) = H_d(q, \mathbf{p}).$$

Since the proof follows *mutatis mutandis*, the proof of Proposition 6.2 details are omitted. As discussed in Section 6.1, to avoid the need to fix the controller initial conditions to ensure the equilibrium assignment, it is necessary to implement the integral action without a dynamic extension. To simplify the presentation, we omit this clarification in the proposition below.

Proposition 7.1: *Consider the underactuated mechanical system* (7.11) *with $M(q)$ and $V(q)$ satisfying Assumptions 7.1–7.3, together with the inner-loop controller* (7.16) *and the PID* (7.22). *Fix a desired assignable equilibrium $(q, \mathbf{p}) = (q^\star, 0)$, with $q_u^\star$ verifying $\nabla V_u(q_u^\star) = 0$. This equilibrium is* globally stable *with Lyapunov function* (7.30)*, where $M_d(q_u)$ and $V_d(q_u)$ are defined in* (7.27) *and* (7.28)*, respectively. The equilibrium is globally* asymptotically *stable if the signal y_d is a* detectable *output for the closed-loop system.*

Remark 7.5: As indicated in the Introduction, similar to the IDA or the CL methods, the system's energy function is shaped with the PID-PBC proposed in this section. However, in contrast with those methods, it is not necessary to solve any PDE, constituting, in this way, a truly constructive controller design method. This feature is achieved abandoning the objective of preservation – in closed loop – of the structure of a mechanical system, either EL or pH, which is the reason why the matching condition of PDEs arise in these methods. Instead, Assumptions 7.1–7.3 are imposed on $M(q)$ and $V(q)$ to assign the desired energy function.

Remark 7.6: It should be pointed out that the $M_d(q)$ and $V_d(q, \mathbf{p})$ that enter in the definition of the Lyapunov function $H_d(q, \mathbf{p})$ given in (7.30) it do not satisfies the matching equations of these methods. However, it has recently been shown in Section 4.1 of (Donaire et al., 2016b) that it satisfies the new matching equations that arise if we add external – nongyroscopic – forces to the desired closed-loop pH system. See equation (21) of (Donaire et al., 2016b).

Remark 7.7: Condition (ii) of Assumption 7.3 – which ensures that $H_d(q, \dot{q})$ is positive definite can be removed by a LaSalle-based convergence analysis under the assumption that the system is *strongly inertially coupled*, that is

$$\text{rank } [m_{au}(q_u)] = s.$$

The details of the proof may be found in Romero et al. (2018).

7.4 PID-PBC of Euler-Lagrange Systems

In order to get a better understanding of the PID-PBC for pH systems derived in Section 7.3, and to make the design of the controller more practical, we translate here the results to the case of EL systems. Furthermore, we establish the relation between the passive outputs defined in Lemma 7.2 and the passive outputs that can easily be obtained for partially linearized EL systems, as is done in Donaire et al. (2016a).

7.4.1 Passive Outputs for Euler–Lagrange Systems

It is well known, see Section 4.5 of van der Schaft (2016) (Spong and Vidyasagar, 2008), and Appendix D.3, that it is possible to go from the pH model of the mechanical system (7.7), with state space (q, p), to the EL representation, with state $(q, \dot{q})$, via the Legendre transformation with the relationship between momenta and velocity being $p = M(q)\dot{q}$. This leads to the EL model

$$M(q)\ddot{q} + C(q, \dot{q})\dot{q} + \nabla V(q) = G(q)u, \tag{7.31}$$

where $C(q, \dot{q})\dot{q}$ are the Coriolis and centrifugal forces.

The total coenergy function of (7.31) is given by

$$H_L(q, \dot{q}) := \frac{1}{2}\dot{q}^\top M(q)\dot{q} + V(q),$$

and it is easy to see that $\dot{H}_L = u^\top G^\top(q)\dot{q}$ – proving that the mapping $u \mapsto G^\top(q)\dot{q}$ is passive. As done in Section 7.3, we make the partition $\dot{q} = \text{col}(\dot{q}_u, \dot{q}_a)$ with $\dot{q}_u \in \mathbb{R}^s$ and $\dot{q}_a \in \mathbb{R}^m$. In addition, we impose the assumption for the input matrix $G(q)$ given by (7.8) and Assumption 7.1.

In Lemma 7.4, we express the passive outputs of Lemma 7.2 – and their associated storage functions – in terms of velocities $\dot{q}$, instead of momenta $\mathbf{p}$, to obtain passive outputs for the EL model (7.31). The interest of this "translation" is that we can implement the PID-PBC in the more accessible coordinates $(q, \dot{q})$ and without the cumbersome change of coordinates introduced in Section 7.3.2. The proof of the lemma follows directly from the identities (7.15) and the equations (7.17), (7.18), and (7.19).

Lemma 7.4: *Consider the pH system* (7.19) *and its associated EL representation* (7.31). *The passive output signals* (7.17) *verify the following:*

$$\begin{aligned} y_u &= -m_{aa}^{-1} m_{au}(q_u)\dot{q}_u \\ y_a &= m_{aa}^{-1} m_{au}(q_u)\dot{q}_u + \dot{q}_a. \end{aligned} \tag{7.32}$$

Furthermore, their associated storage functions (7.18) *maybe written as*

$$\begin{aligned} H_u(q_u, \mathbf{p}_u) &= \dot{q}_u^\top m_{uu}^s(q_u)\dot{q}_u + V_u(q_u) \\ H_a(\mathbf{p}_a) &= \dot{q}^\top M_a(q_u)\dot{q} \end{aligned} \tag{7.33}$$

where

$$\begin{aligned} m_{uu}^s(q_u) &:= m_{uu}(q_u) - m_{au}^\top(q_u) m_{aa}^{-1} m_{au}(q_u) \\ M_a(q_u) &:= \begin{bmatrix} m_{au}^\top(q_u) m_{aa}^{-1} m_{au}(q_u) & m_{au}^\top(q_u) \\ m_{au}(q_u) & m_{aa} \end{bmatrix}. \end{aligned}$$

Remark 7.8: From (7.32), it is clear that $y_u + y_a = \dot{q}_a$, and from (7.33), we see that

$$H_u(q_u, \dot{q}) + H_a(q_u, \dot{q}) = \frac{1}{2}\dot{q}^\top M(q_u)\dot{q} + V_u(q_u),$$

that is, the total coenergy of the system in closed loop with (7.16). In other words, the new passive outputs are obtained splitting the (1, 1) block of the kinetic energy function into two components with one containing the Schur complement of the (2, 2) block $m_{uu}^s(q_u)$.

7.4.2 Passive Outputs for Euler–Lagrange Systems in Spong's Normal Form

In Donaire et al. (2016a), a PID-PBC is designed for a partially linearized EL system, that is for a system in Spong's normal form

$$\begin{aligned} m_{uu}(q_u)\ddot{q}_u + c_u(q_u, \dot{q}_u)\dot{q}_u + \nabla V_u(q_u) &= -m_{au}^\top(q_u)u \\ \ddot{q}_a &= u, \end{aligned} \tag{7.34}$$

with $m_{uu}(q_u)$ verifying the skew-symmetry property

$$\dot{m}_{uu}(q_u) = c_u(q_u, \dot{q}_u) + c_u^\top(q_u, \dot{q}_u).$$

Lemma 7.5 follows trivially from the equations above.

Lemma 7.5: *The signals*

$$\bar{y}_a := \dot{q}_a, \quad \bar{y}_u := -m_{au}(q_u)\dot{q}_u$$

are passive outputs for the system (7.34), *with storage functions*

$$\bar{H}_a(\dot{q}_a) = \frac{1}{2}|\dot{q}_a|^2, \qquad \bar{H}_u(q_u, \dot{q}_u) = \frac{1}{2}\dot{q}_u^\top m_{uu}(q_u)\dot{q}_u + V_u(q_u). \tag{7.35}$$

More precisely,

$$\dot{\bar{H}}_a = u^\top \bar{y}_a, \quad \dot{\bar{H}}_u = u^\top \bar{y}_u.$$

Remark 7.9: Some simple calculations show that the passive outputs of Lemmas 7.4 and 7.5 are related as follows:

$$\begin{bmatrix} \bar{y}_u \\ \bar{y}_a \end{bmatrix} = \begin{bmatrix} m_{aa} & 0 \\ I_m & I_m \end{bmatrix} \begin{bmatrix} y_u \\ y_a \end{bmatrix}.$$

7.5 Extensions

In this section, we present the following extensions of the previous results.

- Extension to the case of tracking constant speed trajectories.
- Proof that the cancellation of the potential energy term $V_a(q_a)$ of the preliminary feedback (7.16) can be obviated for a certain class of potential energy functions.
- Replacement of the *linear* integral term of the PID, that is $K_I x_c$, by a general nonlinear function.

7.5.1 Tracking Constant Speed Trajectories

In this section, it is shown that the controller methodology presented in Proposition 7.1 can be directly extended to track constant speed trajectories in the actuated coordinates with constant positions in the underactuated ones. To formulate this problem, we define the generalized coordinates errors as

$$\tilde{q}(t) = \begin{bmatrix} \tilde{q}_u(t) \\ \tilde{q}_a(t) \end{bmatrix} := \begin{bmatrix} q_u(t) - q_u^\star \\ q_a(t) - rt \end{bmatrix}, \tag{7.36}$$

with $q_u^\star \in \mathbb{R}^s$ and $r \in \mathbb{R}^m$ a *constant* vector. Consistent with the desired trajectories, we define the errors in momenta as

$$\tilde{p} = \begin{bmatrix} \tilde{p}_u \\ \tilde{p}_a \end{bmatrix} := \begin{bmatrix} p_u \\ p_a - T_3^{-1} r \end{bmatrix}. \tag{7.37}$$

The tracking objective is to ensure

$$\lim_{t\to\infty} \begin{bmatrix} \tilde{q}(t) \\ \tilde{p}(t) \end{bmatrix} = 0, \tag{7.38}$$

and the main result, whose proof maybe found in Romero et al. (2016) is as follows:

Proposition 7.2: *Consider the underactuated mechanical system* (7.11) *with $M(q)$ and $V(q)$ satisfying Assumptions 7.1 and 7.2, together with* (7.16) *and the PID control*

$$\begin{aligned} \dot{x}_c &= y_d \\ k_e u &= -K_P \mathbf{y}_d - K_I x_c - K_D \dot{\mathbf{y}}_d, \end{aligned} \tag{7.39}$$

with

$$\mathbf{y}_d := k_a T_3 \tilde{\mathbf{p}}_a + k_u T_2(q_u) \tilde{\mathbf{p}}_u.$$

If Assumption 7.3 is verified, all trajectories of the closed-loop system are bounded, the zero equilibrium of the error system is stable and (7.38) *is satisfied if $\mathbf{y}_d$ is a* detectable *output for the closed-loop system.*

7.5.2 Removing the Cancellation of $V_a(q_a)$

To carry out this extension, the key step for the design of the PID-PBC is to prove that, even without the cancellation of the term $V_a(q_a)$ – that is, without the innerloop (7.16) – the mappings $\tau \mapsto y_u$ and $\tau \mapsto y_a$ are *passive* with suitable storage functions. This fact is stated in the proposition below and requires the following assumption:

Assumption 7.4: The function $V_a(q_a)$ is of the form

$$V_a(q_a) = b_a^\top q_a + c_0, \tag{7.40}$$

with $b_a \in \mathbb{R}^m$ and $c_0 \in \mathbb{R}$.

Lemma 7.6: *Consider the underactuated mechanical system* (7.31) *satisfying Assumptions 7.1, 7.2, and 7.4 and the output signals* (7.32). *The operators $\tau \mapsto y_u$ and $\tau \mapsto y_a$ are* passive *with storage functions*

$$\begin{aligned} \bar{\bar{H}}_u(q_u, \dot{q}_u) &:= \bar{H}_u(q_u, \dot{q}_u) - V_0(q_u) \\ \bar{\bar{H}}_a(q, \dot{q}) &:= \bar{H}_a(\dot{q}_a) - V_a(q_a) + V_0(q_u), \end{aligned} \tag{7.41}$$

where $\bar{H}_a(\dot{q}_a)$ and $\bar{H}_u(q_u, \dot{q}_u)$ are given in (7.35), and we defined

$$V_0(q_u) := b_a^\top V_N(q_u) \tag{7.42}$$

where $V_N(q_u)$ is given in (7.25).

Proof. First, we observe from (7.25) that $\dot{V}_N = y_u$. Also, from (7.32), we have that $y_a = \dot{q}_a - y_u$. Hence, differentiating (7.42), we get

$$\dot{V}_0 = \nabla V_a^\top y_u = \nabla V_a^\top (\dot{q}_a - y_a). \tag{7.43}$$

Now, the time derivative of the storage functions (7.41) yields

$$\dot{\bar{\bar{H}}}_u = \dot{\bar{H}}_u - \dot{V}_0 = (u - \nabla V_a)^\top y_u = \tau^\top y_u$$
$$\dot{\bar{\bar{H}}}_a = \dot{\bar{H}}_a - \dot{V}_a + \dot{V}_0 = u^\top y_a - \nabla V_a^\top \dot{q}_a + \nabla V_a^\top (\dot{q}_a - y_a) = \tau^\top y_a$$

where we used Assumption 7.4, (7.16), (7.42), and (7.43). □

Remark 7.10: In contrast to Lemmata 7.3 and 7.5, Lemma 7.6 does not include any nonlinearity cancellation. On the other hand, we impose Assumption 7.4.

7.5.3 Enlarging the Class of Integral Actions

An additional DOF to verify Assumption 7.4 is obtained replacing the integral term in the controller (7.22) of Proposition 7.1 by

$$k_e u = -K_P y_d - \nabla\Phi(\gamma) - K_D \dot{y}_d,$$

where $\Phi : \mathbb{R}^m \to \mathbb{R}_+$ is an *arbitrary* function verifying $\Phi(\gamma(q^\star)) = 0$ and using the function

$$V_d(q) = k_e k_u V_u(q_u) + \Phi(\gamma(q)),$$

instead of (7.28) in the Lyapunov function (7.30). It is easy to see that the "new" $H_d(q, \mathbf{p})$ still verifies $\dot{H}_d = -\|y_d\|^2_{K_P}$ and all the subsequent analysis follows *verbatim*.

The three extensions above can be combined yielding a robust, higher-performance controller that does not cancel the term $V_a(q_a)$, relaxes Assumption 7.4 and allows for nonlinear integral action – e.g. using a saturation function.

7.6 Examples

In this section, we evaluate the performance of the PID-PBCs proposed in the chapter with two physical examples that have been widely studied in the literature. Namely, tracking of the inverted pendulum on a cart and the cart-pendulum on an inclined plane (Bloch et al., 2001).

7.6.1 Tracking for Inverted Pendulum on a Cart

For this classical single input, 2-DOF example q_u is the angle of the pendulum with respect to the upright vertical position and q_a denotes the position of the cart. The inertia matrix is

$$M(q_u) = \begin{bmatrix} 1 & b\cos(q_u) \\ b\cos(q_u) & m_3 \end{bmatrix},$$

and the potential energy and input vector are

$$V(q_u) = a\cos(q_u); \quad G = e_2,$$

with a, b and m_3 positive parameters.

The objective is to stabilize, with the controller of Proposition 7.2, the up-right vertical position of the pendulum, that is, $q_u^\star = 0$, and impose a *ramp trajectory* to the cart of the form $q_a^\star(t) = rt,\ r \in \mathbb{R}$.

Assumptions 7.1 and 7.2 are clearly satisfied with $V_N(q_u) = b\sin(q_u)$. Now, computing from (7.10) and (7.13) the Cholesky factorization, we get

$$T(q_1) = \begin{bmatrix} \frac{\sqrt{m_3}}{\sqrt{\delta(q_u)}} & 0 \\ -\frac{b\cos(q_u)}{\sqrt{m_3}\sqrt{\delta(q_u)}} & \frac{1}{\sqrt{m_3}} \end{bmatrix},$$

where we defined the function

$$\delta(q_u) := m_3 - b^2\cos^2(q_u) > 0.$$

We now proceed to verify Assumption 7.3. From (7.26), (7.27), and (7.28) and setting $k_a = 1$, it results

$$K(q_u) = k_e + \frac{K_D}{m_3} + k_u K_D \frac{b^2\cos^2(q_u)}{m_3\delta(q_u)}$$

$$M_d^{-1}(q_u) = \begin{bmatrix} k_u^2 K_D \frac{b^2\cos^2(q_u)}{m_3\delta(q_u)} + k_e k_u & -k_u K_D b \frac{\cos(q_u)}{m_3\sqrt{\delta(q_u)}} \\ -k_u K_D b \frac{\cos(q_u)}{m_3\sqrt{\delta(q_u)}} & k_e + \frac{K_D}{m_3} \end{bmatrix}$$

and

$$V_d(q) = ak_e k_u \cos(q_u) + \frac{K_I}{2}\left[q_a + \frac{b(1-k_u)}{m_3}\sin(q_u)\right]^2. \tag{7.44}$$

Some simple calculations show that $\nabla V_d(0) = 0$ for all gains. Now,

$$\nabla^2 V_d(0) = \begin{bmatrix} -ak_e k_u + \frac{b^2(1-k_u)^2}{m_3^2}K_I & \frac{K_I b(1-k_u)}{m_3} \\ \frac{K_I b(1-k_u)}{m_3} & K_I \end{bmatrix}.$$

Hence, to ensure $0 = \arg\min V_d(q)$, it is *necessary and sufficient* that $k_e k_u < 0$. Notice from (7.44) that this corresponds to flipping upside-down the potential energy function $V(q_u)$ – an action that is not possible without the inclusion of the gain k_e premultiplying u in (7.22). From the latter condition and the (1, 1) term of $M_d^{-1}(q_u)$, we see that there is no choice of the gains that will satisfy $M_d(q_u) > 0$ for $|q_u| \geq \frac{\pi}{2}$–therefore, the stability of the desired equilibrium is only local. On the other hand, it is possible to show that, given any $\epsilon > 0$, there exists gains, with $k_e k_u < 0$, such that

$$M_d(q_u) > 0, \ K(q_u) \neq 0, \quad \forall q_u \in \left[\frac{\pi}{2} - \epsilon, \frac{\pi}{2} + \epsilon\right].$$

This guarantees that the domain of attraction of the PID contains the whole (open) half plane for the pendulum.

7.6.2 Cart-Pendulum on an Inclined Plane

In this section, we consider the cart-pendulum on an inclined plane system depicted in Figure 7.10. The objective is to stabilize a desired position of the cart as well as the pendulum at the upright position applying the PID-PBC of Proposition 7.1.

The dynamics of the system has the form (7.31) with $n = 2$, q_a the position of the car, q_u the angle of the pendulum with respect to the up-right vertical position and u a force on the cart. The inertia matrix is

$$M(q_u) = \begin{bmatrix} m\ell^2 & m\ell\cos(q_u - \psi) \\ m\ell\cos(q_u - \psi) & M_c + m \end{bmatrix},$$

with M_c, m the masses of the cart and pendulum, respectively, ℓ the pendulum length and ψ the angle of inclination of the plane. The potential energy function is

$$V(q) = mg\ell\cos(q_u) - (M_c + m)g\sin(\psi)q_a,$$

and the input matrix is $G = \text{col}(0, 1)$. The desired equilibrium is $(0, q_a^\star)$ with $q_a^\star \in \mathbb{R}$, which is the only constant assignable equilibrium point.

This system clearly satisfies Assumptions 7.1 and 7.4 with $s_a = -(M_c + m)g\sin(\psi)$ and $c_0 = 0$. Applying Lemma 7.3, we identify the cyclo–passive outputs as

$$\begin{aligned} y_a &= \dot{q}_a + \frac{m\ell}{M_c + m}\cos(q_u - \psi)\dot{q}_u \\ y_u &= -\frac{m\ell}{M_c + m}\cos(q_u - \psi)\dot{q}_u. \end{aligned}$$

The signal y_d defined in (7.32) takes the form

$$y_d = k_a \dot{q}_a + (k_a - k_u)\frac{m\ell}{M_c + m}\cos(q_u - \psi)\dot{q}_u.$$

In addition, Eq. (7.42) is satisfied with

$$V_N(q_u) = \frac{m\ell}{M_c + m}\sin(q_u - \psi).$$

Finally, the PID controller (7.22) is rewritten as

$$K(q_u)u = -K_p y_d - K_i x_C - S(q, \dot{q})$$

with $x_C = k_a q_a + (k_a - k_u)\frac{m\ell}{M_c+m}\sin(q_u - \psi) + \kappa$, and

$$\begin{aligned}
S(q,\dot{q}) = &- k_u K_D \left\{ -\frac{m\ell}{M_c + m}\sin(q_u - \psi)\dot{q}_u^2 + \mathcal{N}(q_u) \right. \\
&\times \left[-m\ell \sin(q_u - \psi)\dot{q}_u^2 (M_c + m) g \sin(\psi)\right] \\
&\left. + \frac{m^2\ell^2 g}{M_c + m}(m_{uu}^s)^{-1}\cos(q_u - \psi)\sin(q_u) \right\} + k_a K_D g \sin(\psi),
\end{aligned}$$

$$\mathcal{N}(q_u) = \frac{m\cos^2(q_u - \psi)}{(M_c + m)\left[M_c + m - m\cos^2(q_u - \psi)\right]},$$

$$K(q_u) = -\frac{k_u K_D m\left[\cos(q_u - \psi)\right]^2}{(M_c + m)\left(M_c + m\left[\sin(q_u - \psi)\right]^2\right)} + k_e + \frac{k_a K_D}{(M_c + m)},$$

$$\kappa = -k_a q_a^\star + (k_a - k_u)\frac{m\ell\sin(\psi)}{M_c + m}.$$

The parameters and initial conditions used in the simulations have been chosen according to Bloch et al. (2001), and they are given as follows: $m = 0.14\ kg$, $M_c = 0.44\ kg$, $l = 0.215\ m$, $\psi = 20$ degrees, $q(0) = (20\text{deg}, -0.6\text{m})$, and $\dot{q}(0) = 0$. The desired equilibrium is set to $q_a^\star = 0\ m$ for $t \in (0, 5)$ and to $q_a^\star = -0.3\ m$ for $t \in (5, 10)$, with $q_u^\star = 0$ always.

The gains of the PID-PBC are chosen as $K_D = 0.1$, $K_P = 1$, and $K_I = 2$. Using these values, we present three set of simulations, where we change one-by-one the gains k_a, k_u, and k_e, while keeping the PID-PBC gains unaltered – always satisfying Assumption 7.3. Variations of the PID-PBC gains were also considered, but their effect was less informative than changing the gains k_a, k_u, and k_e. In all cases, we present the transient behavior of q, $\dot{q}$, u, and the factor $K(q_u)$. Figures 7.1–7.3 correspond to variations of k_a with $k_u = -500$ and $k_e = 5$. In Figures 7.4–7.6, we change now k_u with $k_a = 50$ and $k_e = 5$. Finally, Figures 7.7–7.9 correspond to variations of k_e

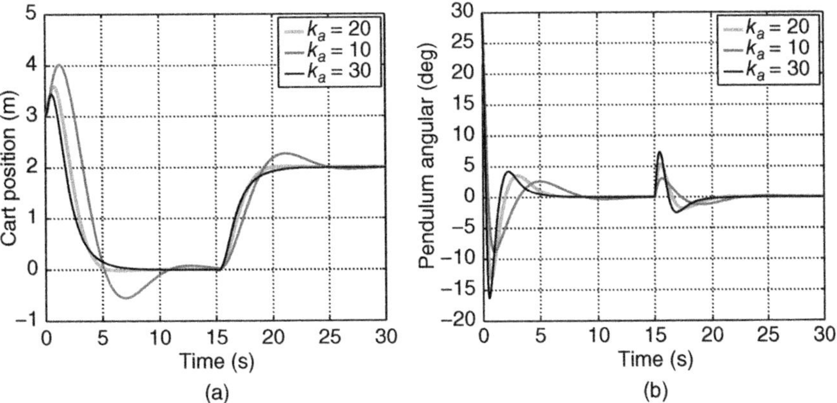

Figure 7.1 Time histories of the (a) position of the cart $q_a(t)$ and (b) angle of the pendulum $q_u(t)$.

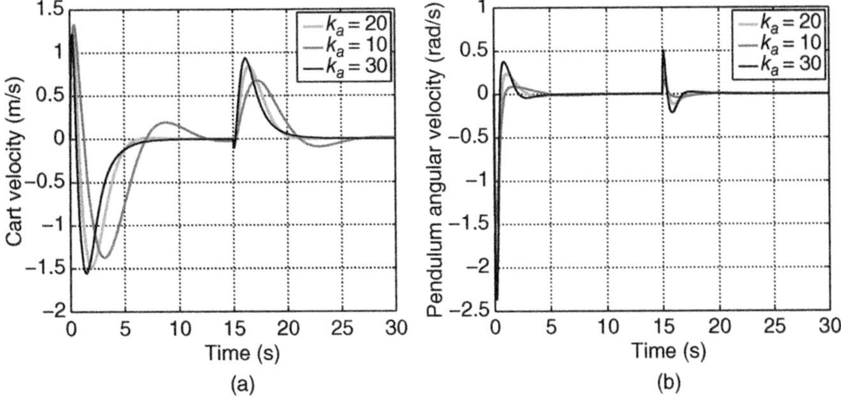

Figure 7.2 Time histories of the (a) velocity of the cart $\dot{q}_a(t)$ and (b) angular velocity of the pendulum $\dot{q}_u(t)$.

with $k_a = 50$ and $k_u = -500$. In all cases, the desired regulation objective is achieved very fast with a reasonable control effort.

These plots should be compared with figure 5 in Bloch et al. (2001), where the transient peaks are much larger, and they take over 100 s to die out. Unfortunately, the plots of the control signal are not shown in Bloch et al. (2001), but given the magnitudes selected in the controller, it is expected to be much larger than the ones resulting from the PID-PBC.

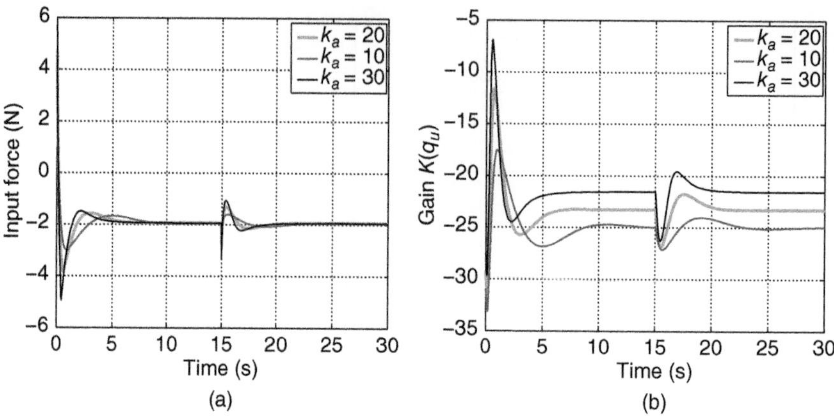

Figure 7.3 Time histories of the (a) input force $u(t)$ and (b) the nonlinear gain $K(q_u)$.

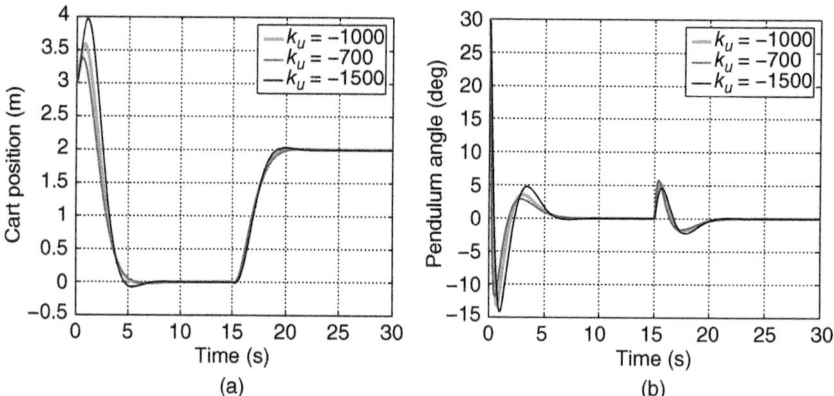

Figure 7.4 Time histories of the (a) position of the cart $q_a(t)$ and (b) angle of the pendulum $q_u(t)$.

Finally, we present in Fig. 7.10 a series of captures of a video animation of the cart-pendulum with initial conditions $(q(0), \dot{q}(0)) = (20\text{deg}, -0.6\text{m})$, $\dot{q}(0) = 0$, and desired equilibrium at the origin. The controller gains were selected as follows: $k_a = 50, k_u = -450, k_e = 5, K_D = 0.1, K_P = 1$, and $K_I = 2$. As it can be seen in the animation, the PID-PBC ensures very good performance while satisfying Assumption 7.3.

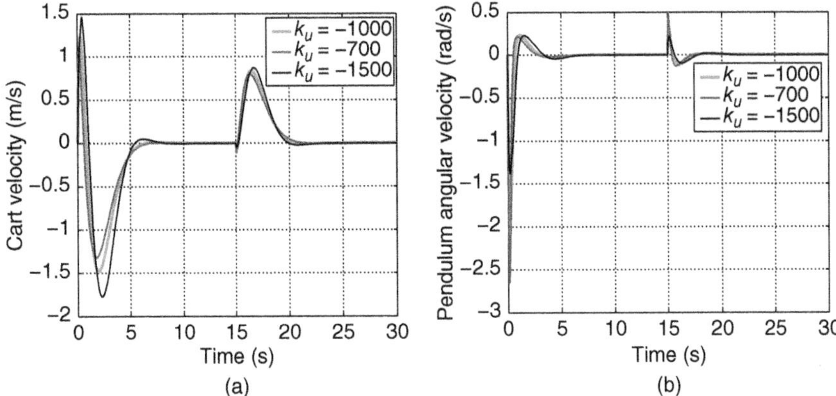

Figure 7.5 Time histories of the (a) velocity of the cart $\dot{q}_a(t)$ and (b) angular velocity of the pendulum $\dot{q}_u(t)$.

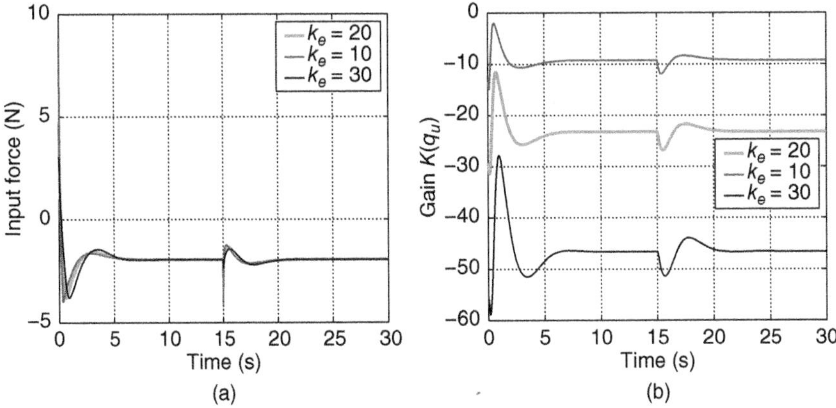

Figure 7.6 Time histories of the (a) input force $u(t)$ and (b) the nonlinear gain $K(q_u)$.

7.7 PID-PBC of Constrained Euler–Lagrange Systems

Control of compliant mechanical systems is increasingly being researched for several applications including flexible link robots and ultraprecision positioning systems. The control problem in these systems is challenging,

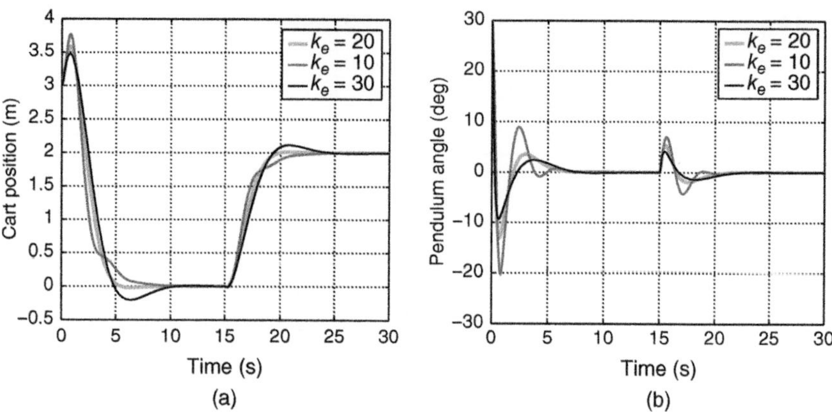

Figure 7.7 Time histories of the (a) position of the cart $q_a(t)$ and (b) angle of the pendulum $q_u(t)$.

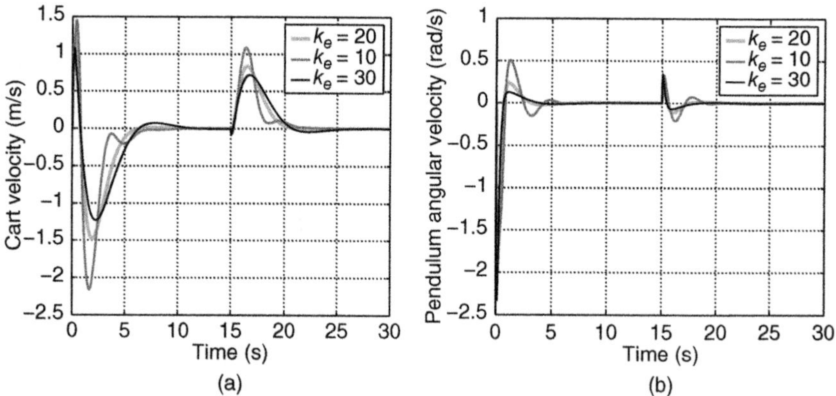

Figure 7.8 Time histories of the (a) velocity of the cart $\dot{q}_a(t)$ and (b) angular velocity of the pendulum $\dot{q}_u(t)$.

especially with gravity coupling and large deformation, because of inherent underactuation and the combination of lumped and distributed parameters of a nonlinear system. If the deformations due to flexibility are small, it is possible to use an *unconstrained* EL formulation and invoke the well-known Assumed Modes Method (AMM) (Meirovitch, 1975) to obtain a finite-dimensional EL model, based on which a controller can be designed. This modeling procedure, however, is inapplicable for systems with large deformations, for which a *constrained* EL formulation is required. The main objective of this chapter is to show that PID-PBC, developed for standard EL

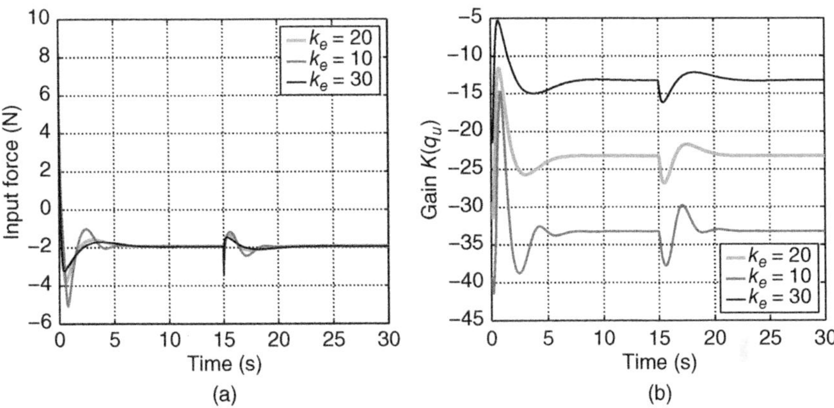

Figure 7.9 Time histories of (a) the input force $u(t)$ and (b) the nonlinear gain $K(q_u)$.

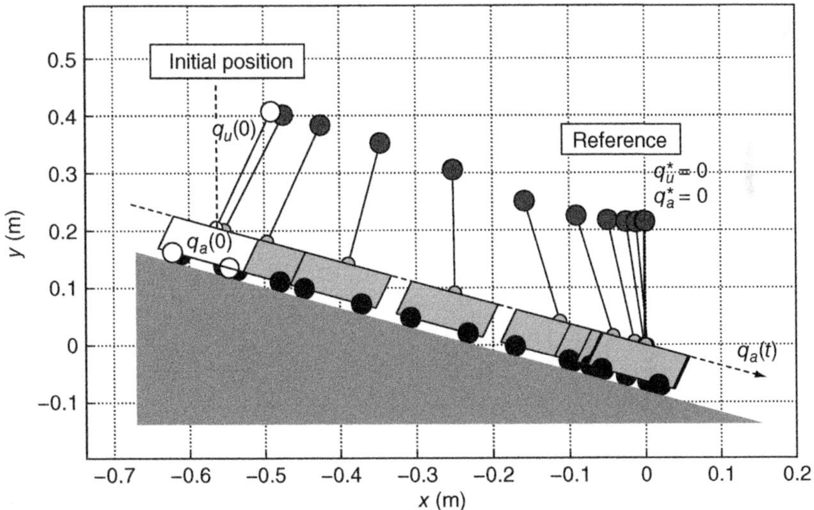

Figure 7.10 Captures of a video animation of the cart-pendulum on an inclined plane. Source: Based on Cart Pendulum Animation. https://youtu.be/CGlnoXkR0FA.

or pH systems, is also applicable to constrained EL systems. This extension is far from obvious because the (lower-order) dynamics that results from the projection of the system on the manifold is defined by the constraint *is not EL*. In spite of this fact, it is shown that it is possible to identify the passive outputs based on which PID-PBC is designed.

To illustrate the extension mentioned above, we consider an ultraflexible inverted pendulum on a cart. The system is modeled using the constrained EL formulation. The use of the constrained EL formalism is necessary because of the ultralarge deformations, and it has been reported in Patil and Gandhi (2014). The potential energy change owing to ultralarge deformations in the presence of gravity is considered in Patil and Gandhi (2014) using the constant length of the beam as a holonomic constraint. Based on this model, which previously has been validated experimentally, we propose a partial feedback linearization step followed by a PID-PBC to bring the pendulum to the upward position with the cart stopped at a desired position. Boundedness of all signals and asymptotic stability of the desired equilibrium is theoretically established. Realistic simulations and experiments assess the performance of the proposed controller.

7.7.1 System Model and Problem Formulation

In this section, we present the model developed by Patil and Gandhi (2014), first, in its constrained EL form and then in a reduced form – obtained via the elimination of the constrained equations. The model admits a constrained EL representation of the form

$$\begin{aligned} D(q)\ddot{q} + C(q,\dot{q})\dot{q} + B(q) + R\dot{q} &= e_3 u + \lambda A(q) \\ \Gamma(q) &= 0, \end{aligned} \tag{7.45}$$

where $q(t) = \mathrm{col}(\theta(t), x_e(t), z(t)) \in \mathbb{D} \times \mathbb{R}_+ \times \mathbb{R}$ are the generalized coordinates, $R \geq 0$ is a matrix of damping coefficients. $D(q) > 0$ is the inertia matrix, $C(q,\dot{q})\dot{q}$ are the Coriolis and centrifugal forces, $B(q)$ is a conservative force vector due to potential energy, $u(t) \in \mathbb{R}$ is the control, λA is a vector of virtual forces due to the holonomic constraint, with λ the Lagrange multiplier, and Γ is the (constant length) constraint function given by

$$\Gamma(q) := \int_0^{x_e} \sqrt{1 + [\theta\phi'(x)]^2}\mathrm{d}x - L, \tag{7.46}$$

with $L > 0$ the length of the link and ϕ the mode shape function of the AMM reported in Laura et al. (1974), that is,

$$\phi(x) = \cosh\left(\frac{\eta x}{L}\right) - \cos\left(\frac{\eta x}{L}\right) + \varrho\left[\sin\left(\frac{\eta x}{L}\right) - \sinh\left(\frac{\eta x}{L}\right)\right],$$

where η and ϱ are positive constants. The analysis made in Patil and Gandhi (2014) considers only one mode where the deflection $\alpha(\theta, x)$ is given by

$$\alpha(x, \theta) = \phi(x)\theta.$$

The different terms entering into (7.45) are defined as

$$D(q) := \begin{bmatrix} D_1(x_e) & 0 & D_2(x_e) \\ 0 & D_3 & 0 \\ D_2(x_e) & 0 & D_4 \end{bmatrix},$$

$$A(q) = \begin{bmatrix} A_1(\theta, x_e) \\ A_2(\theta, x_e) \\ 0 \end{bmatrix} := \nabla\Gamma(q), \;\; R := \mathrm{diag}\{R_1, 0, R_3\},$$

$$C(q, \dot{q}) := \begin{bmatrix} \frac{1}{2}C_1(x_e)\dot{x}_e & \delta(x_e, \dot{\theta}, \dot{z}) & \frac{1}{2}C_2(x_e)\dot{x}_e \\ -\delta(x_e, \dot{\theta}, \dot{z}) & 0 & -\frac{1}{2}C_2(x_e)\dot{\theta} \\ \frac{1}{2}C_2(x_e)\dot{x}_e & \frac{1}{2}C_2(x_e)\dot{\theta} & 0 \end{bmatrix},$$

with

$$\delta(x_e, \dot{\theta}, \dot{z}) = \frac{1}{2}C_1(x_e)\dot{\theta} + \frac{1}{2}C_2(x_e)\dot{z},$$

and

$$B(q) = \begin{bmatrix} B_1(\theta, x_e) \\ B_2(\theta, x_e) \\ 0 \end{bmatrix} := \begin{bmatrix} \frac{\partial V(q)}{\partial \theta} \\ \frac{\partial V(q)}{\partial x_e} \\ 0 \end{bmatrix}, \tag{7.47}$$

where $V(q)$ is the potential energy of the system given by

$$V(q) = \frac{1}{2}EI\int_0^{x_e} \frac{[\theta\phi''(x)]^2}{\{1 + [\theta\phi'(x)]^2\}^3}dx - D_3 g(L - x_e),$$

E, I, D_3, D_4, R_1, R_3 are positive constants and the remaining functions may be found in Gandhi et al. (2016).

Problem Formulation

Given the system (7.45), find control input u that places the beam at its vertical position with the cart stopped at the zero position, i.e. that asymptotically stabilizes the point $q^\star := (0, L, 0)$ (Figure 7.11).

Remark 7.11: In Patil and Gandhi (2014), the analysis of the open-loop equilibria of (7.45) is carried out. In particular, it is proven that the open-loop equilibrium set is given by

$$\begin{aligned} \mathcal{E} := \{&(\theta, x_e, z) \in \mathbb{D} \times \mathbb{R}_+ \times \mathbb{R} \mid \\ &A_1(\theta, x_e)B_2(\theta, x_e) - A_2(\theta, x_e)B_1(\theta, x_e) = 0\}. \end{aligned} \tag{7.48}$$

Furthermore, and not surprisingly, it is shown that the desired equilibrium $q^\star \in \mathcal{E}$ and is unstable.

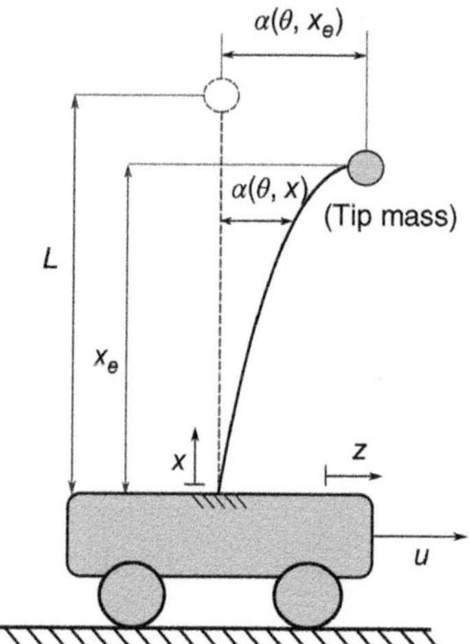

Figure 7.11 Single ultraflexible link with base excitation.

7.7.2 Reduced Purely Differential Model

In this section, we apply the standard constraint differentiation procedure (Hill and Mareels, 1990) to transform the differential-algebraic equations (7.45) to a purely differential form of reduced order.

Proposition 7.3: *The system dynamics* (7.45) *is equivalent to*

$$\begin{aligned} D_\theta(\theta)\ddot{\theta} + D_z(\theta)\ddot{z} + C_\theta(\theta)\dot{\theta}^2 + R_1\dot{\theta} + B_\theta(\theta) = 0 \\ D_z(\theta)\ddot{\theta} + D_4\ddot{z} + C_z(\theta)\dot{\theta}^2 + R_3\dot{z} = u, \end{aligned} \tag{7.49}$$

with

$$\begin{aligned} D_\theta(\theta) &:= D_1(\hat{x}_e(\theta)) + D_3\frac{A_1^2(\theta,\hat{x}_e(\theta))}{A_2^2(\theta,\hat{x}_e(\theta))} \\ C_\theta(\theta) &:= D_3\frac{A_1(\theta,\hat{x}_e(\theta))}{A_2^2(\theta,\hat{x}_e(\theta))}\zeta - \frac{1}{2}C_1(\hat{x}_e(\theta))\frac{A_1(\theta,\hat{x}_e(\theta))}{A_2(\theta,\hat{x}_e(\theta))} \\ B_\theta(\theta) &:= B_1(\theta,\hat{x}_e(\theta)) - B_2(\theta,\hat{x}_e(\theta))\frac{A_1(\theta,\hat{x}_e(\theta))}{A_2(\theta,\hat{x}_e(\theta))} \\ D_z(\theta) &:= D_2(\hat{x}_e(\theta)) \\ C_z(\theta) &:= -C_2(\hat{x}_e(\theta))\frac{A_1(\theta,\hat{x}_e(\theta))}{A_2(\theta,\hat{x}_e(\theta))}, \end{aligned}$$

where

$$\zeta = A_5(\theta,\hat{x}_e(\theta)) + A_4(\theta,\hat{x}_e(\theta))\frac{A_1^2(\theta,\hat{x}_e(\theta))}{A_2^2(\theta,\hat{x}_e(\theta))} - A_3(\theta,\hat{x}_e(\theta))\frac{A_1(\theta,\hat{x}_e(\theta))}{A_2(\theta,\hat{x}_e(\theta))}.$$

Remark 7.12: Since $\Gamma(q) = 0$, then $\dot{\Gamma}(q) = A^\top(q)\dot{q} = 0$. Consequently, differentiating the total energy of (7.45) and using the well-known skew-symmetry property yields, the usual power-balance equation $\dot{H} = -\dot{q}^\top R\dot{q} + \dot{q}_3 u$. This means that the virtual forces introduced in the equations due to constrained Lagrange formulation are workless, that is, they are not responsible for addition or removal of energy from the system. This key property is used later to identify the passive outputs used for the design of the energy shaping controller.

Remark 7.13: The admissible initial conditions for the reduced system (7.49) are restricted to the set $\{(\theta, z) \in \mathbb{D} \times \mathbb{R} \mid \Gamma(\theta, \hat{x}_e(\theta)) = 0\}$, where the system evolves.

7.7.3 Design of the PID-PBC

As explained in the introduction, PID-PBC proceeds in three steps; first a static state-feedback that performs the partial feedback linearization of the system (7.49), then the identification of the passive output and, finally, the design of the PID-PBC. The three steps are summarized below.

Lemma 7.7: *Consider the system* (7.49) *in closed-loop with the control*

$$u = R_3\dot{z} + \left(C_z - \frac{D_z}{D_\theta}C_\theta\right)\dot{\theta}^2 - \frac{D_z}{D_\theta}R_1\dot{\theta} - \frac{D_z}{D_\theta}B_\theta + \left(D_4 - \frac{D_z^2}{D_\theta}\right)v. \tag{7.50}$$

Then, the partially linearized system can be written in Spong's normal form

$$\begin{aligned} D_\theta(\theta)\ddot{\theta} + C_\theta(\theta)\dot{\theta}^2 + R_1\dot{\theta} + B_\theta(\theta) &= G_\theta(\theta)v \\ \ddot{z} &= v, \end{aligned} \tag{7.51}$$

where $G_\theta(\theta) := -D_z(\theta)$.

Lemma 7.8: *Consider the system given in* (7.51). *The signals*

$$y_a := \dot{z},\ y_u := G_\theta(\theta)\dot{\theta},$$

define cyclo-passive maps $v \mapsto y_a$ *and* $u \mapsto y_u$ *with storage functions*

$$H_a(z) = \frac{1}{2}\dot{z}^2 \tag{7.52}$$

$$H_u(\theta) = \frac{1}{2}D_\theta(\theta)\dot{\theta}^2 + V_\theta(\theta), \tag{7.53}$$

respectively, where $V_\theta(\theta) := V(\theta, \hat{x}_e(\theta))$. *More precisely,* $\dot{H}_a \leq v y_a$, $\dot{H}_u \leq v y_u$.

The controller design is completed adding a PID around a suitably weighted sum of the two cyclo-passive outputs (y_a and y_u) identified in Lemma 7.5, that is

$$\begin{aligned} \dot{x}_c &= y_d \\ k_e v &= -K_P y_d - K_I x_c - K_D \dot{y}, \end{aligned} \tag{7.54}$$

where $y_d := k_a y_a + k_u y_u$, with $k_e, k_a, k_u \in \mathbb{R}$ and $K_P, K_I, K_D \in \mathbb{R}_+$. As explained in Section 7.4, and illustrated below, these gains are selected to shape the energy function.

To implement the controller (7.54) *without differentiation*, the term $\dot{y}_d$ is replaced by its evaluation along the system dynamics. Since the system is relative degree one, this brings along some terms depending on v that are moved to the left hand side of (7.54). Some lengthy, but straightforward, calculations show that (7.54) is equivalent to

$$K(\theta)u = -K_P y_d - K_I x_c - K_D k_u S(\theta, \dot{\theta}),$$

where we defined the functions

$$\begin{aligned} S(\theta, \dot{\theta}) &:= \dot{G}_\theta \dot{\theta} - \frac{G_\theta}{D_\theta}\left[C_\theta \dot{\theta}^2 + R_1 \dot{\theta} + B_\theta\right] \\ K(\theta) &:= k_e + K_D \left[k_a + k_u \frac{G_\theta^2(\theta)}{D_\theta(\theta)}\right]. \end{aligned}$$

Clearly, a sufficient condition for the controller to be implementable is that $|K(\theta)| \geq \delta > 0$.

To analyze the stability of the PID (7.54), we propose the function

$$U(x_c, \tilde{y}, \theta, \dot{\theta}, \dot{z}) := k_e[k_a H_a(\dot{z}) + k_u H_u(\theta, \dot{\theta})] + \frac{1}{2}(K_I x_c^2 + K_D y_d^2),$$

whose derivative along the trajectories of the system (7.51) in closed loop with (7.54) yields $\dot{U} = -K_P \tilde{y}^2$, where we *neglected* the friction term $R_1 \dot{\theta}^2$.

As usual in PID-PBC, the final step in our stability analysis is to express x_c as a function of the system state. For, we define the function

$$V_N(\theta) := -D_3 \int_0^\theta \phi(\hat{x}_e(s))\mathrm{d}s - \left(\rho A_0 \int_0^L \phi(x)\mathrm{d}x\right)\theta,$$

whose time derivative is given by

$$\dot{V}_N = -D_3 \phi(\hat{x}_e(\theta))\dot{\theta} - \left(\rho A_0 \int_0^L \phi(x)\mathrm{d}x\right)\dot{\theta} = G_\theta \dot{\theta} = y_u. \tag{7.55}$$

Consequently,

$$x_c = k_a z(t) + k_u V_N(\theta(t)) + \kappa =: \gamma(z, \theta)$$

where $\kappa \in \mathbb{R}$ is an integration constant. Using the latter and the definitions of $H_a(z), H_u(\theta)$, and y_d, we can prove that, up to an additive constant,

$$U(\gamma(z,\theta), y_d, \theta, \dot{\theta}, \dot{z}) = \frac{1}{2}\begin{bmatrix}\dot{\theta}\\ \dot{z}\end{bmatrix}^{\top} D_d(\theta)\begin{bmatrix}\dot{\theta}\\ \dot{z}\end{bmatrix} + V_d(\theta, z) =: H_d(\theta, z, \dot{\theta}, \dot{z}) \tag{7.56}$$

where we defined

$$D_d(\theta) := \begin{bmatrix} k_e k_u D_\theta(\theta) + k_u^2 K_D G_\theta^2(\theta) & k_a k_u K_D G_\theta(\theta) \\ k_a k_u K_D G_\theta(\theta) & k_e k_a + k_a^2 K_D \end{bmatrix} \tag{7.57}$$

$$V_d(\theta, z) := k_e k_u V_\theta(\theta) + \frac{1}{2} K_I \left[k_a z + k_u V_N(\theta)\right]^2. \tag{7.58}$$

7.7.4 Main Stability Result

The proposition below, which essentially gives conditions on the controller gains to ensure H_d is positive definite, is the main result of this section, whose proof may be found in Gandhi et al. (2016).

Proposition 7.4: *Consider the system* (7.45) *in closed loop with the controller* (7.50) *and the PID-PBC* (7.54) *with*

$$y_d = k_a \dot{z} - k_u D_z(\theta)\dot{\theta}. \tag{7.59}$$

Set the constant gains k_e, k_a and the PID gains K_P, K_I and K_D to arbitrary positive *numbers, while k_u is negative and, for some small $\epsilon > 0$, satisfies*

$$k_u \leq -C\left(k_a + \frac{k_e}{K_D}\right) - \epsilon, \tag{7.60}$$

where C is a positive constant verifying $C \geq \frac{D_\theta(\theta)}{G_\theta^2(\theta)}$.

The origin of the reduced dynamics (7.49), *which corresponds to the desired equilibrium $q^\star = (0, L, 0)$ of* (7.45), *is* stable *with Lyapunov function H_d given in* (7.56). *It is* asymptotically *stable if the signal y_d defined in* (7.59) *is detectable with respect to* (7.51).

Remark 7.14: Notice that $\nabla^2 V_\theta(0) < 0$, which is consistent with the well-known fact that the upward pendulum position is unstable in open loop. Similar to the rigid case treated in Section 7.7, the maximum of the open-loop potential energy is transformed into a minimum in closed-loop multiplying V_θ by the negative number $k_e k_u$ – see (7.58).

Remark 7.15: The critical condition (7.60) is satisfied in a neighborhood of the origin replacing C by

$$\frac{D_3\phi^2(L) + \rho\mathcal{A}_0 \int_0^L \phi^2(x)dx}{\left[D_3\phi(L) + \rho\mathcal{A}_0 \int_0^L \phi(x)dx\right]^2} = \frac{D_\theta(0)}{G_\theta^2(0)}.$$

Remark 7.16: The term $k_a z + k_u V_N(\theta)$ in (7.58) is a new potential energy corresponding to a virtual spring attached to the cart – thereby enabling to stabilize the cart position.

7.7.5 Simulation Results

In this section, we assess the performance of the proposed controller via Matlab simulations choosing different sets of gains and different initial conditions. We simulated the system (7.51) in closed-loop with the PID controller (7.54) with the parameters given in Table 7.1.

We have chosen three different initial conditions, given in Table 7.2, corresponding to radically different scenarios of the system. Namely, an arbitrary

Table 7.1 System parameters.

Parameter	Symbol	Value
Pendulum cross-section area	$\mathcal{A}_0$	$8 \times 10^{-6}\ \mathrm{m}^2$
Young's modulus	E	$9 \times 10^{10}\ \frac{\mathrm{N}}{\mathrm{m}^2}$
Gravitational acceleration	g	$9.81\ \frac{\mathrm{m}}{\mathrm{seg}^2}$
Moment of inertia	I	$1.066 \times 10^{-13}\ \mathrm{kg} \cdot \mathrm{m}^2$
Pendulum length	L	$0.305\ \mathrm{m}$
Tip mass	M	$2.75 \times 10^{-2}\ \mathrm{kg}$
Cart mass	M_c	$0.1\ \mathrm{kg}$
Function of the system natural frequency	η	1.1741
Dimensionless constant	ϱ	0.9049
Pendulum density	ρ	$8400\ \frac{\mathrm{kg}}{\mathrm{m}^3}$
Viscous friction at the pendulum base	R_1	$9.86 \times 10^{-4}\ \frac{\mathrm{kg}}{\mathrm{seg}}$
Viscous friction between the rail and the cart	R_3	$7.69\ \frac{\mathrm{kg}}{\mathrm{seg}}$

Table 7.2 Initial conditions.

	θ (m)	z (m)	$\dot{\theta}$ (m/s)	$\dot{z}$ (m/s)
$\mathcal{A}p_{s1}$	−0.08	−0.1	0	0
$\mathcal{A}p_{s2}$	0.134	0	0	0
$\mathcal{A}p_{s3}$	0	−0.15	0	0

Table 7.3 Gains sets.

	k_e	k_a	k_u	K_D	K_P	K_I
Set 1	1	0.5	−50.77	1.47	1.94	0.35
Set 2	1	1	−61.37	1.28	1.92	0.52
Set 3	1	1	−43.04	2.18	3.66	1.35

point ($\mathcal{A}p_{s1}$), one of the stable open-loop equilibria ($\mathcal{A}p_{s2}$), and an initial condition with the cart far from the origin and the tip mass located at the unstable open-loop equilibrium ($\mathcal{A}p_{s3}$).

For the selection of suitable gains for the controller, we fixed the gain $k_e = 1$ and linearized the closed-loop system. We based our criterion to choose the gains on the eigenvalues of the closed-loop matrix of the linearized system around the desired equilibrium point. Particular attention has been paid to the eigenvalue closest to the imaginary axis, which is directly related to the rate of convergence of the cart position. Three sets of gains were selected, and they are given in Table 7.3. For the Set 1, the real part of the slowest pole was −0.58, −0.75 for the Set 2 and −1.33 for Set 3.

Simulation results of the PID-PBC are shown in Figures 7.12–7.14, where the variation of the cart position and control input acceleration is observed to be within practical limits; hence, the control objective of simultaneous stabilization of cart position while suppressing the cantilever vibrations is achieved.

7.7.6 Experimental Results

Experiments were also carried out to assess the performance of the proposed controller. Figure 7.15 shows the picture of the setup used for experimental implementation. A fatigue-resistant Cu–Be alloy material is used for fabrication of the beam. Cart is guided by a rail and driven through a toothed belt driven by a motor (Maxon Motor AG: 236670). An encoder reads the position

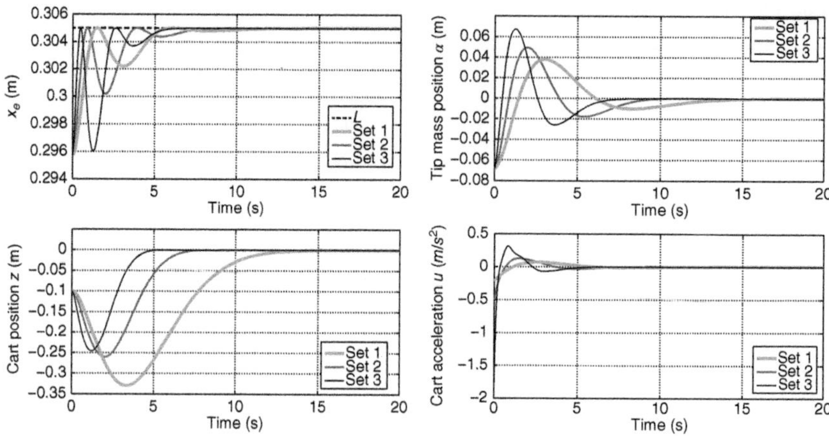

Figure 7.12 Simulation results for $\mathcal{A}p_{s1}$.

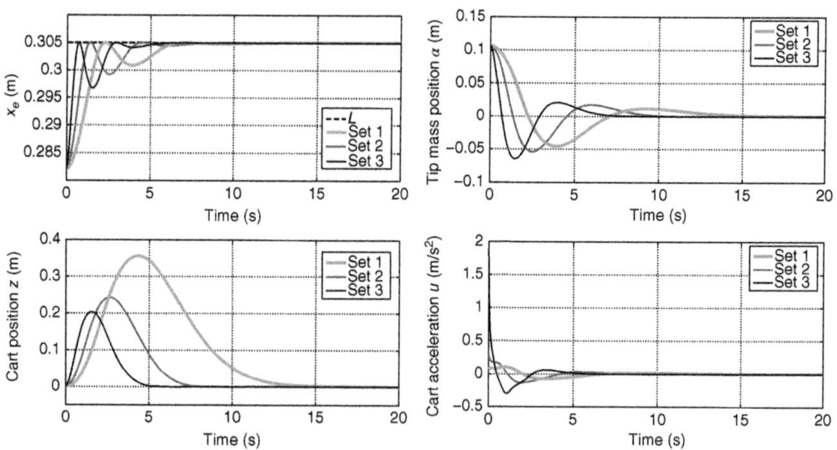

Figure 7.13 Simulation results for $\mathcal{A}p_{s2}$.

z of the motor and hence the cart. An H-bridge amplifier (Nex Robotics Hercules 36V,15A) is used to drive the motor. Strain gauges (TML Tokyo Sokki Kenkyujo Co.: FLA-5-11) in full bridge configuration along with an amplifier (DataQ Instruments 5B38-02) are used for feedback θ. The derivatives $\dot{\theta}$ and $\dot{z}$ are computed numerically using a digital derivative filter. Interfacing of the motor, strain amplifier, and encoder is done with data acquisition system ds 1104 from dSPACE GmbH via PWM, digital-to-analog converter (DAC), and encoder interfaces. Careful horizontal leveling of the cart and rail, and meticulous adjustment of the beam and the center of gravity of the tip mass

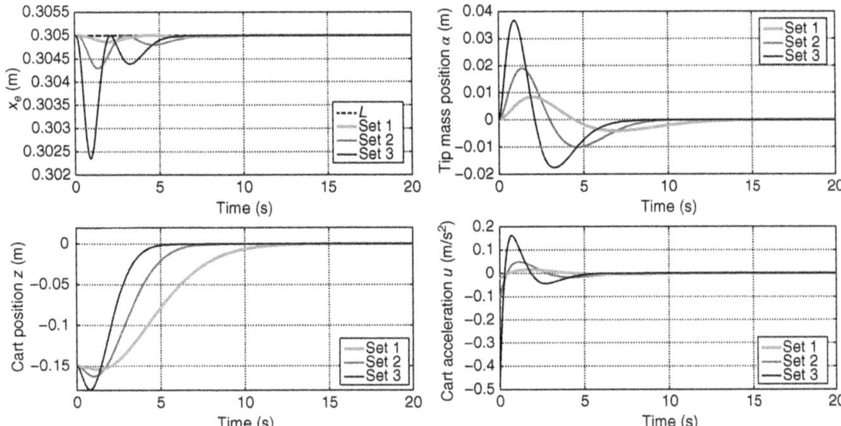

Figure 7.14 Simulation results for $\mathcal{A}p_{s3}$.

Figure 7.15 Inverted flexible pendulum.

is carried out to make sure that the unstable equilibrium is perfectly vertical and other equilibria are symmetric about the unstable equilibrium position. Several nonlinear terms in the control law are integral function of θ with length constraint giving $\dot{x}_e$ as limit of integration and thus are computationally demanding to evaluate in real time. Hence, a look-up table arrangement is used for evaluation of these terms in real time. Appropriate signal conditioning is used to balance detrimental effects of noise on one side and filter delay on the other.

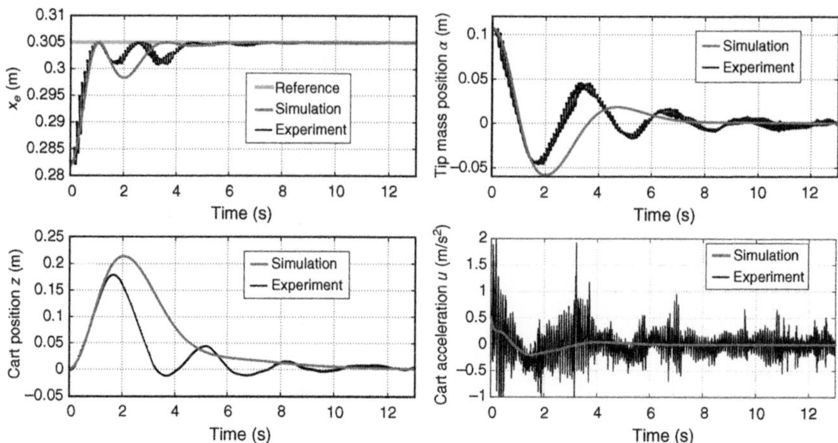

Figure 7.16 Comparison of simulation and experimental results for $\mathcal{A}p_{s2}$.

It was observed that the gains used in the simulation do stabilize the physical pendulum but with a reduced domain of attraction, that is, placing the pendulum closer to the upward position. It is not surprising that the domain of attraction predicted by the model is reduced in the practical application. To show a comparison of the simulation and the experiment starting from the same initial conditions and using the same controller, it was decided to select another set of gains. Figure 7.16 presents the comparison of simulation and experiment for the set of initial conditions $\mathcal{A}p_{s2}$ and the set of gains: $k_e = 1$, $k_a = 1$, $k_u = -47.5$, $K_D = 1.9$, $K_P = 3$, and $K_I = 0.9$. The results demonstrate that the control task is achieved in a similar time although the trajectory in the experiments shows more oscillations with high-frequency components of the vibrations of the beam. These oscillations are not captured by the simulation model that, as mentioned retains only the first deflection mode. However, as shown in the plots, these high-frequency vibrations degrade the transient performance but do not induce instability. A video of the experiments can be watched at https://youtu.be/aG53XaQPP3c.

Bibliography

A. Ailon and R. Ortega. An observer-based controller for robot manipulators with flexible joints. *Systems & Control Letters*, 21: 329–335, 1993.

N. Bedrossian and M. W. Spong. Feedback linearization of robot manipulators and Riemannian curvature. *Journal of Robotic Systems*, 12(8): 541–552, 1995.

G. Blankenstein, R. Ortega, and A. J. van der Schaft. The matching conditions of controlled Lagrangians and IDA-passivity based control. *International Journal of Control*, 75(9): 645–665, 2002.

A. M. Bloch, D. E. Chang, N. E. Leonard, and J. E. Marsden. Controlled Lagrangians and the stabilization of mechanical systems II: potential shaping. *IEEE Transactions on Automatic Control*, 46(10): 1556–1571, 2001.

A. M. Bloch, N. E. Leonard, and J. E. Marsden. Controlled Lagrangians and the stabilization of mechanical systems. *IEEE Transactions on Automatic Control*, 45(12): 2253–2270, 2000.

A. Donaire, R. Mehra, R. Ortega, S. Satpute, J. G. Romero, F. Kazi, and N. M. Singh. Shaping the energy of mechanical systems without solving partial differential equations. *IEEE Transactions on Automatic Control*, 61(4): 1051–1056, 2016a.

A. Donaire, R. Ortega, and J. G. Romero. Simultaneous interconnection and damping assignment passivity-based control of mechanical systems using dissipative forces. *Systems & Control Letters*, 94: 118–126, 2016b.

L. P. Eisenhart. *Riemannian Geometry. Princeton Landmarks in Mathematics and Physics*, Princeton University Press, Princeton, NJ, 1997.

K. Fujimoto and T. Sugie. Canonical transformation and stabilization of generalized Hamiltonian systems. *Systems & Control Letters*, 42(3): 217–227, 2001.

P. Gandhi, P. Borja, and R. Ortega. Energy shaping control of an inverted flexible pendulum fixed to a cart. *Control Engineering Practice*, 56(11): 27–36, 2016.

D. J. Hill and I. Mareels. Stability theory for differential/algebraic systems with application to power systems. *IEEE Transactions on Circuits and Systems*, 37(11): 327–357, 1990.

E. Jonckheere. Lagrangian theory of large scale systems. In *European Conference on Circuit Theory and Design*, pages 626–629, The Hague, The Netherlands, 1981.

H. Khalil. *Nonlinear Systems*. Prentice-Hall, Upper Saddle River, NJ, 2002.

D. E. Koditschek. Robot planning and control via potential functions. *The Robotics Review I*, In: Khatib, O., Craig, Khatib, O., Craig (eds.). The MIT Press, Cambridge: 349–367, 1989.

P. Laura, J. L. Pombo, and E. Susemihl. A note on the vibrations of a clamped-free beam with a mass at the free end. *Journal of Sound and Vibration*, 37(2): 161–168, 1974.

R. Mehra, S. G. Satpute, F. Kazi, and N. M. Singh. Control of a class of underactuated mechanical systems obviating matching conditions. *Automatica*, 86: 98–103, 2017.

L. Meirovitch. *Elements of Vibration Analysis*. McGraw-Hill, 1975.

H. Nijmeijer and A. J. van der Schaft. *Nonlinear Dynamical Control Systems.* Springer-Verlag, London, 1990.

R. Ortega, A. Donaire, and J. G. Romero. *Passivity–Based Control of Mechanical Systems*, pages 167–199. Feedback Stabilization of Controlled Dynamical Systems–In Honor of Laurent Praly, Lecture Notes in Control and Information Sciences. Springer-Verlag, Berlin/Heidelberg, 2017.

R. Ortega, J. A. Loría, P. J. Nicklasson, and H. Sira-Ramírez. *Passivity-Based Control of Euler-Lagrange Systems.* Springer-Verlag, 1998.

R. Ortega, M. W. Spong, F. Gómez-Estern, and G. Blankenstein. Stabilization of a class of underactuated mechanical systems via interconnection and damping assignment. *IEEE Transactions on Automatic Control*, 47(8): 1218–1233, 2002a.

R. Ortega, A. J. van der Schaft, B. M. Maschke, and G. Escobar. Interconnection and damping assignment passivity–based control of port–controlled Hamiltonian systems. *Automatica*, 38(4): 585–596, 2002b.

O. Patil and P. Gandhi. On the dynamics and multiple equilibria of an inverted flexible pendulum with tip mass on a cart. *Journal of Dynamic Systems, Measurement, and Control*, 136(4): 041017 (9 pages), 2014.

J. G. Romero, A. Donaire, R. Ortega, and P. Borja. Global stabilisation of underactuated mechanical systems via PID passivity-based control. *Automatica*, 96(10): 178–185, 2018.

J. G. Romero, R. Ortega, and A. Donaire. Energy shaping of mechanical systems via PID control and extension to constant speed tracking. *IEEE Transactions on Automatic Control*, 61(11): 3551–3556, 2016.

A. Shiriaev, L. Freidovich, and M. W. Spong. Controlled invariants and trajectory planning for underactuated mechanical systems. *IEEE Transactions on Automatic Control*, 59(9): 2555–2561, 2014.

M. W. Spong. Partial feedback linearization of underactuated mechanical systems. In *IEEE/RSJ/GI International Conference on Intelligent Robots and Systems*, Munich, Germany, 1994.

M. W. Spong and M. Vidyasagar. *Robot Dynamics and Control.* John Wiley & Sons, Inc., New York, 2008.

M. Takegaki and S. Arimoto. A new feedback method for dynamic control of manipulator. *ASME Journal of Dynamics Systems, Measurement, and Control*, 103: 119–125, 1981.

A. J. van der Schaft. *L_2-Gain and Passivity Techniques in Nonlinear Control.* Springer-Verlag, Berlin, 3rd edition, 2016.

A. Venkatraman, R. Ortega, I. Sarras, and A. J. van der Schaft. Speed observation and position feedback stabilization of partially linearizable mechanical systems. *IEEE Transactions on Automatic Control*, 55(5): 1059–1074, 2010.

8

Disturbance Rejection in Port-Hamiltonian Systems

We have discussed in Chapters 4, 6 and 7, the application of PID-PBC for stabilization of nonlinear systems, with particular emphasis on pH systems. It is well known that, due to its energy-shaping nature, PBC is robust with respect to parametric uncertainty and passive unmodeled dynamics (like friction in mechanical systems), in the sense that stability – with respect to a shifted equilibrium – is preserved. However, in practical applications, measurement or system noise is unavoidable, and it is usually necessary to robustify the controller with respect to this *external, additive disturbances*, which were neglected in Chapters 6 and 7.

In this chapter, we explicitly take into account the presence of these disturbances and propose various methods to attenuate their effect on the performance of the closed-loop system – in all cases, the problem is solved via the addition of an *integral action*.[1] The scenario we consider is as follows: given a nonlinear system that, in the absence of external disturbances has a stable equilibrium at the desired operating point $x^\star$, design an additional IA loop such that "some stability properties" are preserved in spite of the presence of external disturbances.

In view of Proposition D.2, without loss of generality, we assume that the system is pH. More precisely, we consider perturbed pH systems of the form

$$\dot{x} = F(x)\nabla H(x) + g(x)u + d, \tag{8.1}$$

where $H(x)$ verifies

$$x^\star = \arg\min \{H(x)\} \tag{8.2}$$

1 As will become clear below the use of the name "integral action" is done with some abuse of notation. But, in all cases, it involves the open-loop integration of a signal.

PID Passivity-Based Control of Nonlinear Systems with Applications, First Edition.
Romeo Ortega, José Guadalupe Romero, Pablo Borja, and Alejandro Donaire.

Published 2021 by John Wiley & Sons, Inc.

and $d \in \mathbb{R}^n$ is an external disturbance. If the disturbance is *constant*, we would like the new closed-loop system to have a globally stable equilibrium, with some of the components of x at their desired equilibrium. If they are *time-varying*, we want to ensure input-to-state-stability (Sontag, 2008) with respect to the external disturbance. Regarding the disturbance we distinguish two cases: when it enters in the imagen of the input matrix – referred as *matched* disturbances – and when the disturbance is *unmatched*. As explained in Section 8.1, in the presence of unmatched disturbances, it is not possible to preserve all the coordinates of the minimizer of the energy function as part of the closed-loop equilibrium.

Two alternative approaches to solve the problem have been reported in the literature. First, the inclusion of a change of coordinates and an IA that transforms the system into some desired pH from, that was first reported in Donaire and Junco (2009). A central feature of this form is that the energy function of the closed loop, in the new coordinates, is the sum of the energy of the original pH system and the energy of the controller and qualifies as a Lyapunov function for the stability analysis. A second approach, first introduced in Ferguson et al. (2017a), avoids the coordinate change, instead, it proposes an energy function for the controller that depends on the state of the plant. This is similar to the CbI used for total energy shaping in Ailon and Ortega (1993).

In the first part of the chapter, we treat general pH systems and, later on, we specialize in the practically important case of mechanical systems.

8.1 Some Remarks on Notation and Assignable Equilibria

8.1.1 Notational Simplifications

As done throughout the book, in this chapter, we have decided to sacrifice generality for clarity of presentation. Consequently, two assumptions that, without modifying the essence of our contribution, considerably simplify the notation are made. First, since we consider the case where disturbances enter in the $s := (n - m)$ nonactuated coordinates, the internal model principle indicates that it is necessary to add s integrators. To ensure solvability of the problem, it is reasonable to assume that the number of control actions is sufficiently large. This leads to the following assumption:

$$m \geq s. \tag{8.3}$$

If less states are added this restriction can be relaxed – but then the notation gets very cumbersome.

The second simplification that we introduce concerns the matrix $g(x)$. Dragging this matrix through the calculations significantly complicates the notation; therefore, it will be assumed in the sequel that, after a change of coordinates, the input matrix takes the form

$$g(x) = \begin{bmatrix} I_m \\ 0 \end{bmatrix}. \tag{8.4}$$

As discussed in Appendix C.2, a necessary and sufficient condition to achieve this transformation is that the columns of the matrix $g(x)$ commute.

For notational convenience, we partition the state and disturbance vectors as $x = \text{col}(x_a, x_u)$, $d = \text{col}(d_a, d_u)$, where $d_a, x_a \in \mathbb{R}^m$ and $d_u, x_u \in \mathbb{R}^s$. Similarly, the matrix $F(x)$ is block partitioned as

$$F(x) = \begin{bmatrix} F_{aa}(x) & F_{au}(x) \\ F_{ua}(x) & F_{uu}(x) \end{bmatrix}, \tag{8.5}$$

with $F_{aa}(x) \in \mathbb{R}^{m\times m}$ and $F_{uu}(x) \in \mathbb{R}^{s\times s}$. With this notation the natural passive output is $y_0 = \nabla_{x_a} H(x)$. For future reference, we also define a second output to be regulated as the $(n-m)$-dimensional vector

$$r = \nabla_{x_u} H(x). \tag{8.6}$$

8.1.2 Assignable Equilibria for Constant *d*

To simplify the presentation, in the sequel, we identify the set of minimizers of $H(x)$ with

$$\mathcal{M} := \{x \in \mathbb{R}^n \mid \nabla H(x) = 0,\ \nabla^2 H(x) > 0\}. \tag{8.7}$$

Since the second-order (Hessian positivity) condition is sufficient, but not necessary, for $x^\star$ to be a minimizer of $H(x)$, the set $\mathcal{M}$ is a subset of the minimizer set; hence, the consideration is taken with a slight loss of generality.

In the perturbed case, the set of assignable equilibria of (8.1), (8.4) is given by

$$\mathcal{E} := \{x \in \mathbb{R}^n \mid F_{ua}(x)\nabla_{x_a} H(x) + F_{uu}(x)\nabla_{x_u} H(x) = -d_u\}. \tag{8.8}$$

It is clear that, if the disturbances are matched, i.e. $d_u = 0$,

$$\mathcal{M} \subseteq \mathcal{E}.$$

That is, all energy minimizers are assignable equilibria, and it is desirable to preserve in closed loop the open-loop equilibria. On the other hand, in the face of unmatched disturbances, that is, when $d_u \neq 0$,

$$\mathcal{M} \cap \mathcal{E} = \emptyset. \tag{8.9}$$

In other words, it is not possible to assign as equilibrium a minimizer of the energy function. As will become clear below, this situation complicates the task of rejection of unmatched disturbances.

8.2 Integral Action on the Passive Output

In the following proposition, the output regulation and disturbance rejection properties of IA on the passive output of a pH system are revisited. Although both properties are widely referred in the literature, for the sake of completeness, we give the result below, which was reported for the first time in Ortega and García-Canseco (2004).

Proposition 8.1: *Consider the perturbed pH system*

$$\begin{aligned} \dot{x} &= F(x)\nabla H(x) + g(x)(u + d_a) \\ y_0 &= g^\top(x)\nabla H(x) \end{aligned} \tag{8.10}$$

verifying (8.2), *in closed-loop with the IA*

$$\begin{aligned} \dot{x}_c &= y_0 \\ u &= -K_I x_c, \end{aligned} \tag{8.11}$$

where $K_I > 0$ *is an arbitrary tuning gain.*

(i) The closed-loop is a pH system of the form

$$\begin{bmatrix} \dot{x} \\ \dot{x}_c \end{bmatrix} = \begin{bmatrix} F(x) & -g(x)K_I \\ K_I g^\top(x) & 0 \end{bmatrix} \nabla W(x,\eta), \tag{8.12}$$

with

$$W(x, x_c) := H(x) + \frac{1}{2}\|x_c - d_a\|^2_{K_I}. \tag{8.13}$$

(ii) The equilibrium $(x^\star, d_a)$ *is stable.*

(iii) There exists a (closed) ball, centered in $(x^\star, d_a)$ *such that for all initial states* $(x(0), x_c(0)) \in \mathbb{R}^n \times \mathbb{R}^m$ *inside the ball, the trajectories are bounded and* $\lim_{t\to\infty} y_0(t) = 0$.

(iv) If, moreover, y_0 *is a* detectable *output for the closed-loop system* (8.10), (8.11), *the equilibrium is asymptotically stable.*

The properties (ii)–(iv) are global if $H(x)$ *is positive definite and radially unbounded.*

The following remarks are in order:

R1 The disturbance is *matched*, that is, it enters in the image of the input matrix $g(x)$, and there was no need to assume that it is of the form (8.4).

R2 The integral control only ensures that $y_0(t) \to 0$ and an additional detectability requirement is needed to ensure $x(t) \to x^\star$. It is, precisely, this restrictive condition that makes the result of little practical interest, and we will propose in the sequel an alternative IA that will obviate it.

R3 An essential step to carry out the stability analysis is the preservation of the pH structure. It is shown in Donaire and Junco (2009) that this feature is lost if we add the IA to an output whose relative degree is larger than one, for instance the signal r defined in (8.6).

R4 Looking at the linearization of the closed-loop system (8.4), (8.10), and (8.11), it is possible to show that, if $x^\star \in \mathcal{M}$ and the (uu) block of the matrix $F(x)$, evaluated at $x^\star$ is full rank, $x^\star$ is an exponentially stable equilibrium. The rank condition holds if and only if the triple

$$\left(F^\star, \begin{bmatrix} -K_I \\ 0 \end{bmatrix}, \begin{bmatrix} K_I & 0 \end{bmatrix}\right)$$

has no transmission zeros at the origin. This assumption is standard for integral control of nonlinear systems. See, e.g. Section 12.3 of (Khalil, 2002).

8.3 Solution Using Coordinate Changes

As discussed above, an essential feature of the pH representation is that it naturally suggests the use of the Hamiltonian function as a Lyapunov function. Motivated by this fact, in Donaire and Junco (2009) posed the problem of finding a coordinate transformation and an IA such that the closed-loop system in the new coordinates – even in the presence of *unmatched* disturbances – takes a form similar to (8.12) with a separable Hamiltonian function like (8.13). More precisely, the new closed-loop should not modify the functional relations in the matrix $F(x)$ nor the energy function $H(x)$.[2] Preserving the energy function avoids the need to solve a PDE, replacing it instead by some algebraic equations, keeping the same interconnection and damping matrix, simplifies these equations. This idea was later generalized in Ortega and Romero (2012) and expressed as a problem of feedback equivalence. In this section, we review the main results for disturbance rejection of pH systems using this approach.

2 See Remark 8.1 for a clarification of this point.

8.3.1 A Feedback Equivalence Problem

To formalize the objective discussed above, we introduce the following definition of feedback equivalence.

Definition 8.1 The perturbed pH system

$$\dot{x} = F(x)\nabla H(x) + \begin{bmatrix} I_m \\ 0 \end{bmatrix} u + \begin{bmatrix} 0 \\ d_u \end{bmatrix} \tag{8.14}$$

is said to be feedback equivalent to a *matched disturbance integral controlled system* – for short, MDICS equivalent – if there exists two mappings

$$\hat{u}, \psi \; : \; \mathbb{R}^m \times \mathbb{R}^s \times \mathbb{R}^s \to \mathbb{R}^m,$$

with

$$\text{rank} \; \{\nabla_{x_a} \psi(x_a, x_u, x_c)\} = m, \tag{8.15}$$

such that the system in closed loop with the "integral" control:

$$\begin{aligned} \dot{x}_c &= K_I[\nabla_{x_u} H(\psi(x_a, x_u, x_c), x_u)] \\ u &= \hat{u}(x_a, x_u, x_c), \end{aligned} \tag{8.16}$$

expressed in the coordinates,

$$\begin{aligned} z_1 &= \psi(x_a, x_u, x_c) \\ z_2 &= x_u \\ z_3 &= x_c, \end{aligned} \tag{8.17}$$

takes the pH-form

$$\dot{z} = \begin{bmatrix} F(z_1, z_2) & \begin{bmatrix} 0 \\ -K_I \end{bmatrix} \\ \begin{bmatrix} 0 & K_I \end{bmatrix} & 0 \end{bmatrix} \nabla U(z), \tag{8.18}$$

where

$$U(z) := H(z_1, z_2) + \frac{1}{2} || z_3 - d_u ||^2_{K_I^{-1}}. \tag{8.19}$$

It is said to be robustly MDICS equivalent if the mappings $\psi(x_u, x_a, x_c)$ and $\hat{u}(x_u, x_a, x_c)$ can be computed without knowledge of the unmatched disturbance d_u.

MDICS equivalence guarantees that the transformed closed-loop system takes the desired form (8.18). The rank condition (8.15) ensures that (8.17) is a diffeomorphism that maps the set of equilibria of the (x, x_c)-system into the equilibria of the z-system. This is, of course, necessary to be able to infer stability of one system from stability of the other one. Robust MDICS equivalence guarantees that the control law (8.16) can be implemented without the knowledge of the unmatched disturbance d_u.

At this point, we make the important observation that choosing the desired value for z_3 to be equal to d_u is necessary to be able to solve the robust MDICS equivalence problem. Indeed, since in the change of coordinates (8.17) we fixed $z_2 = x_u$, and these are unactuated coordinates, it is necessary that d_u, which appears in $\dot{x}_u$, appears also in $\dot{z}_2$. Remark that, since $z_3 = x_c$, the equilibrium value for x_c is also d_u.

As explained in Section 8.1.2, the equilibrium sets of (8.14), (8.16), and (8.18) are not just different, but they are actually disjoint, see (8.9). Indeed, while the (x components of the) former are in the set

$$\mathcal{E}_{cl} := \mathcal{E} \cap \{x \in \mathbb{R}^n \mid [\nabla_{x_u} H](\psi(x_a, x_u, d_u), x_u) = 0\}, \tag{8.20}$$

the (z_1, z_2) components of the latter are in $\mathcal{M}$. In spite of that, the fact that (8.17) is a diffeomorphism ensures that the implication

$$(x_a, x_u) \in \mathcal{E}_{cl} \ \Rightarrow \ (\psi(x_a, x_u, d_u), x_u) \in \mathcal{M}, \tag{8.21}$$

is true, which will be essential for future developments.

Remark 8.1: It is important to underscore that in the feedback equivalence problem considered here the matrix $F(z_1, z_2)$, and energy function $H(z_1, z_2)$ are just the evaluations of the original functions of the x system in the z coordinates, without applying the (inverse) change of coordinates.[3] That is, $H(z_1, z_2)$ is not obtained from the composition of $H(x_1, x_2)$ with the inverse of the change of coordinates (8.17), but is simply $H(z_1, z_2) = H(x_a, x_u)|_{x_a = z_1, x_u = z_2}$. This, rather arbitrary, choice is done to be able to translate MDICS equivalence into an *algebraic* problem.

Remark 8.2: The proposed control (8.16) is, in general, not an integral action because of the possible dependence of $\psi(x_u, x_a, x_c)$ with respect to x_c. We have decided to keep the name because in the z coordinates it is, indeed, an integral action of the form

$$\dot{z}_3 = K_I \nabla_{z_2} H(z_1, z_2). \tag{8.22}$$

3 To avoid cluttering the notation the same symbols, $H(\cdot)$ and $F(\cdot)$, have been used for both functions.

8.3.2 Local Solutions of the Feedback Equivalent Problem

In Proposition 2 of (Ortega and Romero, 2012), it is shown that the *global* solution of the MDICS problem boils down to solution of s algebraic equations of the form $L_i(x_a, x_u, x_c, \psi(x_a, x_u, x_c)) = 0,\ i \in \bar{s} = \{1, \cdots, s\}$, that needs to be solved for the mapping $\psi(x_a, x_u, x_c)$. The robust MDICS requirement impose a constraint on these mapping of the form $\nabla_{x_u}\psi(x_a, x_u, x_c)d_u = 0$.

In the proposition below, we give conditions for the *local* solution of the problem. The proof of this proposition may be found in Ortega and Romero (2012). To streamline the presentation of the result, define the linearization of the pH system (8.14) at the points $x^\star \in \mathcal{E}_{cl}$ and $\bar{x} \in \mathcal{M}$ as

$$A := \nabla(F(x)\nabla H(x))|_{x=x^\star}, \quad E := (F(x)\nabla^2 H(x))|_{x=\bar{x}}. \tag{8.23}$$

These $n \times n$ matrices are block partitioned as

$$A = \begin{bmatrix} A_{aa} & A_{au} \\ A_{ua} & A_{uu} \end{bmatrix},$$

with $A_{aa} \in \mathbb{R}^{m\times m}$ and $A_{uu} \in \mathbb{R}^{s\times s}$, with a similar partition for E. Notice that, since $\nabla H(\bar{x}) = 0$, the linearization at a point in the minimizer set takes a simpler form.

Proposition 8.2: *Consider the perturbed pH system* (8.14) *satisfying condition* (8.3) *and two points: $x^\star \in \mathcal{E}_{cl}$ and $\bar{x} \in \mathcal{M}$.*

(NC) *A necessary condition for MDICS equivalence is that the linearizations of the pH system at the points $x^\star$ and $\bar{x}$ are controllable. That is, the pairs*

$$\left(A, \begin{bmatrix} I_m \\ 0 \end{bmatrix}\right), \left(E, \begin{bmatrix} I_m \\ 0 \end{bmatrix}\right)$$

are controllable pairs.

(SC) *A sufficient condition for MDICS equivalence is that the* (ua) *blocks of the matrices A and E defined in* (8.23) *are full rank. That is,*

$$\text{rank}\,\{A_{ua}\} = \text{rank}\,\{E_{ua}\} = s. \tag{8.24}$$

Moreover, the system is robustly MDICS equivalent if

$$\begin{aligned} A_{uu} &= E_{uu} \\ A_{ua}x_u^\star &= -d_u. \end{aligned} \tag{8.25}$$

Remark 8.3: Unfortunately, there is a gap between the necessary and the sufficient conditions of Proposition 8.2. Indeed, controllability of the linearized systems is necessary, but not sufficient, for MDICS equivalence. The gap stems from the fact that, without further qualifications on E_{uu}, controllability does not ensure that rank $\{E_{ua}\} = s$. On the other hand, it is obvious that (8.24) implies controllability.

8.3.3 Stability of the Closed-Loop

The following proposition summarizes the properties of MDICS equivalent systems. More precisely, the IA preserves stability of the equilibrium, and it is robust *vis-á-vis* unmatched disturbance d_u.

Proposition 8.3: *Consider the perturbed pH system* (8.14) *satisfying condition* (8.3)*. Assume there exist two points, $x^\star \in \mathcal{E}_{cl}$ and $\bar{x} \in \mathcal{M}$, that is, an assignable equilibrium and a minimizer of the energy $H(x)$, such that* (8.24) *and* (8.25) *hold, with A and E defined in* (8.23)*. Under these conditions, there exist two mappings*

$$\hat{u}, \psi : \mathbb{R}^m \times \mathbb{R}^s \times \mathbb{R}^s \rightarrow \mathbb{R}^m,$$

such that the "integral" control (8.16) *ensures the following properties:*

(i) ***Stability of the equilibrium***. *The equilibrium $(x_1^\star, x_2^\star, d_u)$ is stable.*

(ii) ***Regulation of the passive output***. *There exists a (closed) ball, centered at the equilibrium, such that for all initial states $(x(0), x_c(0)) \in \mathbb{R}^n \times \mathbb{R}^s$ inside the ball the trajectories are bounded and*

$$\lim_{t\to\infty} y_0(t) = 0.$$

(iii) ***Asymptotic stability***. *If, moreover, y_0 is a detectable output for the closed-loop system* (8.14), (8.16)*, the equilibrium is asymptotically stable.*

(iv) ***Regulation of the nonpassive output***. *Under the condition (iii), there exists a (closed) ball, centered at the equilibrium, such that for all initial states $(x(0), x_c(0)) \in \mathbb{R}^n \times \mathbb{R}^s$ inside the ball the trajectories are bounded and the output* (8.6) *satisfies*

$$\lim_{t\to\infty} r(t) = 0.$$

Remark 8.4: In Donaire and Junco (2009), the IA proposed in this chapter is applied to reject the torque disturbances – that are unmatched – of PMSMs. In Proposition 5 of (Ortega and Romero, 2012), it is shown that fully damped LTI pH systems with the block F_{au} *full rank* are robustly MDICS equivalent. Hence, unmatched disturbances can be rejected with the IA of Definition 8.1. A class of mechanical systems that also satisfy this condition are discussed in Section 8.8.1.

8.4 Solution Using Nonseparable Energy Functions

The Hamiltonian of the IA discussed above depends only on the integrator coordinates and proposes – in the spirit of standard CbI (van der Schaft, 2016) – a *separable* Lyapunov function consisting of the sum of this plant and the controller energy functions. As proposed in Ailon and Ortega (1993), it is also possible to propose controller energy functions that explicitly depend on the state of the plan, see Nuño and Ortega (2018) for a recent application of this idea in the context of multiagent systems.

This key modification was first applied for disturbance rejection of pH systems in Ferguson et al. (2017a). It was later extended in Ferguson et al. (2020) for the particular case of matched disturbances removing the stringent constraint of detectability needed to ensure asymptotic stability in the classical scheme of Proposition 8.1 and incorporating some additional robustness features. In this section, we review the main results obtained using this approach that has the significant advantage of avoiding the need of the cumbersome coordinate change of the previous method.

8.4.1 Matched and Unmatched Disturbances

In Ferguson et al. (2017a) pH systems of the form (8.1), (8.2) with the partitions (8.4), (8.5) and the following assumptions are considered.

C1 $F_{ua}(x) = -F_{au}^{\top}(x)$. That is, the damping matrix is block diagonal.
C2 $F_{aa}(x) + F_{aa}^{\top}(x)$ is full rank. That is, the actuated part of the dynamics is fully damped.
C3 $H(x) = H_a(x_a) + H_u(x_u)$ and is strongly convex.

The control objective is to find an IA such that the equilibrium $\bar{x} = (\bar{x}_a, x_u^{\star}, \bar{x}_c)$, for some constant $\bar{x}_a \in \mathbb{R}^m$ and $\bar{x}_c \in s$, is GAS in spite of the presence of constant d. That is, it is desired to preserve the desired equilibrium for the unactuated coordinate.

The following IA is proposed

$$\begin{aligned} \dot{x}_c &= E^{\top} F_{au}(x) \nabla_{x_u} H(x) \\ u &= F_{aa}(x) \nabla_{x_a} H_c(E^{\top} x_a - x_c) \end{aligned} \tag{8.26}$$

where $x_c \in \mathbb{R}^s$ and $H_c(\cdot)$ is a strictly convex function of $z := E^{\top} x_a - x_c$ such that ∇H_c is invertible.

The main motivation to consider this kind of control is that the closed-loop dynamics preserves the pH structure and is of the form

$$\begin{bmatrix} \dot{x} \\ \dot{x}_c \end{bmatrix} = \begin{bmatrix} F(x) & \begin{bmatrix} 0 \\ -F_{au}^\top(x)E \end{bmatrix} \\ \begin{bmatrix} 0 & E^\top F_{au}(x) \end{bmatrix} & 0 \end{bmatrix} \nabla H_{cl}(x, x_c) - \begin{bmatrix} d_a \\ d_u \\ 0 \end{bmatrix}, \tag{8.27}$$

where $H_{cl}(x, x_c) := H(x) + H_c(E^\top x_a - x_c)$. It is important to underscore that

$$\begin{bmatrix} \nabla_{x_a} H(x) \\ \nabla_{x_u} H(x) \\ \nabla_z H_c(z) \end{bmatrix} = \begin{bmatrix} \nabla_{x_a} H_{cl} + E^\top \nabla_{x_c} H_c \\ \nabla_{x_u} H_{cl} \\ -\nabla_{x_c} H_{cl} \end{bmatrix}, \tag{8.28}$$

which are instrumental to carry out the stability analysis of the closed-loop system (8.27), presented in the sequel. The proofs of these propositions may be found in Ferguson et al. (2017a).

Matched Disturbances

Consider the case of matched disturbances only, that is d_a is an unknown constant and $d_u = 0$, together with the following assumptions:

C4 F_{aa} is constant.

C5 There exists a constant matrix E such that $E^\top F_{au}(x)$ is invertible and

$$[F_{au}^\top(\overline{x})E]^{-1}F_{au}^\top(\overline{x}) = [F_{au}^\top(x)E]^{-1}F_{au}^\top(x).$$

C6 There exists an unique $\overline{x}_a$ such that

$$\nabla_{x_a} H(\overline{x}) = [I_m - E(J_{12}^\top(\overline{x})E)^{-1}J_{12}^\top(\overline{x})]\; F_{aa}^{-1}(x)\; d_a.$$

Remark 8.5: Notice that **C5** is for two important cases: (i) if F_{au} is invertible, then any invertible matrix E will satisfied the assumption and (ii) if F_{au} is constant, then any constant matrix E such that $E^\top F_{au}$ is invertible will satisfy the assumption, e.g. $E = F_{au}$.

Remark 8.6: Assumption **C6** is satisfied if F_{au} is invertible. In this case $\nabla_{x_a} H(\overline{x}) = 0$, which is verified for $\overline{x}_a = x_a^\star$.

Proposition 8.4: *Consider system* (8.1) *subject to unknown, matched disturbance in closed loop with the controller* (8.26)*. Then, under Assumptions* ***C1–C6****, the equilibrium of the closed loop* $\overline{x} = (\overline{x}_a, x_u^\star, \overline{x}_c)$ *is GAS.*

Unmatched Disturbances

Consider now the case of unmatched disturbances only, that is $d_a = 0$ and d_u is an unknown constant, together with the following assumptions:

C7 F_{au} is constant.
C8 There exists an unique $\overline{x}_a$ such that $\nabla_{x_a} H(\overline{x}) = -E(J_{12}^{\top}(\overline{x})E)^{-1} d_u$.

Proposition 8.5: *Consider system* (8.1) *subject to unmatched constant disturbance in closed loop with the controller* (8.26)*. Then, under Assumptions* ***C1–C3, C7*** *and* ***C8****, the equilibrium of the closed loop* $\overline{x} = (\overline{x}_a, x_u^{\star}, \overline{x}_c)$ *is GAS.*

Matched and Unmatched Disturbances

In this section, we consider the both matched and unmatched disturbances, that is d_a and d_u are unknown, constant disturbances, and the following additional assumption:

C9 There exists an unique $\overline{x}_a$ such that

$$\nabla_{x_a} H(\overline{x}) = [I_m - E(J_{12}^{\top}(\overline{x})E)^{-1} J_{12}^{\top}(\overline{x})]\, F_{aa}^{-1}(x)\, d_a - E(J_{12}^{\top}(\overline{x})E)^{-1} d_u.$$

Proposition 8.6: *Consider system* (8.1) *subject to both matched and unmatched constant disturbances in closed loop with the controller* (8.26)*. Then, under Assumptions* ***C1–C4****,* ***C7****, and* ***C9****, the equilibrium of the closed loop* $\overline{x} = (\overline{x}_a, x_u^{\star}, \overline{x}_c)$ *is GAS.*

8.4.2 Robust Matched Disturbance Rejection

In this section, we present a solution to the problem of rejection only of matched disturbance, which has the following significant benefits:

B1 Compared with the IA for passive output regulation explored in Section 8.1.1, the restrictive detectability requirement is significantly relaxed by preservation of the stability properties of the unperturbed open-loop system.
B2 Compared with the IA using coordinate transformation explored in Section 8.1.2 and (Donaire and Junco, 2009; Ortega and Romero, 2012): (i) the need to solve a nonlinear algebraic equation is obviated; (ii) the construction of the closed-loop Hamiltonian function is more natural in the sense that the open-loop energy function is unaltered through the construction process; and (iii) the resulting IAs are considerably simpler.
B3 Compared with the IA without coordinate transformation explored in Ferguson et al. (2015, 2017b) and Ferguson et al. (2018): (i) the assumptions of convexity and separability of the systems Hamiltonian

H are removed; (ii) the detectability assumption required for asymptotic stability is relaxed; and (iii) the proposed scheme is shown to be robust against a wider class of modeling uncertainty than has been previously considered.

Decomposition of the Damping Matrix

Since the matrix $F(x) + F^{\top}(x)$ is symmetric positive semidefinite and assume that its rank is k, there exists a unitary matrix $U(x) \in \mathbb{R}^{n\times n}$ such that

$$F(x) + F^{\top}(x) = U(x)\mathrm{diag}\{\lambda_1(x), \dots, \lambda_k(x), \underbrace{0, \dots, 0}_{n-k}\}U^{\top}(x) \tag{8.29}$$

where λ_j, $j = 1, \dots, k$, are the positive eigenvalues of $F + F^{\top}$. Consequently, partitioning the unitary matrix U by

$$U(x) = \begin{bmatrix} U_a(x) & U_u(x) \end{bmatrix}, \tag{8.30}$$

where $U_a(x) \in \mathbb{R}^{n\times k}$ and $U_u(x) \in \mathbb{R}^{n\times(n-k)}$. This partition allows to write $\dot{H}$ as

$$\dot{H} = -\nabla^{\top}H\ U_a\ \mathrm{diag}\{\lambda_1, \dots, \lambda_k\}\ U_a^{\top}\ \nabla H. \tag{8.31}$$

Thus, it follows that $\dot{H} = 0 \ \Rightarrow\ U_a^{\top}\nabla H = 0$.

Related to the decomposition (8.29), a necessary assumption for our control scheme is required in order to ensure robustness against modeling error. Let $\hat{F}$ be an estimate for F, and define

$$\tilde{F}(x) := F(x) - \hat{F}(x) \tag{8.32}$$

as the modeling error associated with the matrix F. We then consider the following assumptions:

C10 The matrix $\tilde{F}(x)$ defined in (8.32) satisfies:

(a) There exists some constant $\kappa \geq 0$ such that

$$\|\tilde{F}(x)\| \leq \kappa \tag{8.33}$$

in the domain of interest.

(b) The columns of $U_u(x)$, defined in (8.30), are contained within the kernel of $\tilde{F}$. Equivalently,

$$\tilde{F}(x)U_u(x) = 0. \tag{8.34}$$

C11 The equilibrium $x^{\star}$ of the undisturbed open-loop system $\dot{x} = F(x)\nabla H$ is asymptotically stable.

Remark 8.7: As the columns of $U_u(x)$ form a basis for the kernel of $F + F^\top$, the condition (8.34) indicates that no modeling errors are permitted in the directions with no damping. It is precisely the damping that is exploited to compensate for the modeling error.

Control Objective

Consider the pH system (8.1), (8.2), (8.4) with $d = \text{col}(d_a, 0)$ verifying assumption **C10**. Design a dynamic state-feedback controller

$$\dot{x}_c = w(x, x_c)$$
$$u = \hat{u}(x, x_c)$$

with $x_c \in \mathbb{R}^m$ and the mapping $w : \mathbb{R}^{n\times m} \mapsto \mathbb{R}^m$, which ensures the closed-loop is an unperturbed pH system with an (asymptotically) stable equilibrium at $(x^\star, x_c^\star)$, for some $x_c^\star \in \mathbb{R}^m$. Consistent with the modeling errors considered in assumption **C10**, the control law should only depend on the estimated system matrix $\hat{F}(x)$.

New Closed-Loop pH Structure

The following proposition presents a novel IA that does not rely on changes of coordinates nor assumes separability of the systems Hamiltonian.

Proposition 8.7: *Consider the pH system* (8.1), (8.2), (8.4), *with* $d = \text{col}(d_a, 0)$, *in closed-loop with the IA controller*

$$\begin{aligned} \dot{x}_c &= g^\top \left(\beta\hat{F} - \alpha I_n\right) \nabla H, \\ u &= -\alpha K_i \left(\beta g^\top x - x_c\right), \end{aligned} \tag{8.35}$$

where $\alpha, \beta \in \mathbb{R}^+$ *and* $K_i = K_i^\top > 0$ *are tuning parameters. Then, defining the new coordinate*

$$z := \beta g^\top x - x_c, \tag{8.36}$$

the closed-loop system can be written in pH form

$$\begin{bmatrix} \dot{x} \\ \dot{z} \end{bmatrix} = \underbrace{\begin{bmatrix} F & -\alpha g \\ \beta g^\top \tilde{F} + \alpha g^\top & -\alpha\beta I_m \end{bmatrix}}_{:=F_{cl}} \begin{bmatrix} \nabla H \\ \nabla H_c \end{bmatrix} \tag{8.37}$$

where $\tilde{F}$ *is the modeling error matrix, defined in* (8.32), *and the controller energy* $H_c(z)$ *is defined by* $H_c(z) := \frac{1}{2}\left\| z - \frac{1}{\alpha}K_i^{-1} d_a \right\|^2_{K_i}$.

The system (8.37) is not necessarily pH. Indeed, due to the presence of the modeling error $\tilde{F}$, the symmetric component of F_{cl} may not be nonpositive. In the sequel, the tuning parameter α is chosen sufficiently large to ensure that the closed-loop system (8.37) is indeed pH.

Stability

Now, the stability properties of the closed-loop system are examined. In the case that the open-loop system is perfectly known ($\hat{F} = F$), stability can be easily verified by taking $H + H_c$ as a Lyapunov candidate. The difficulty in this proof is to verify stability in the case the F is not exactly known.

Proposition 8.8: *The following claims are true for the closed-loop system* (8.37)*:*

(i) The system has an equilibrium point

$$(x, x_c) = \left(x^\star, \underbrace{\beta G^\top x^\star + \frac{1}{\alpha} K_i^{-1} d}_{:= x_c^\star} \right). \tag{8.38}$$

(ii) If the error matrix $\tilde{F}$ *satisfies assumption* ***C10****, then the equilibrium* (8.38) *is stable provided that the control gains* α *and* β *satisfy*

$$\frac{\alpha}{\beta} > \frac{\kappa^2}{4 \, \min\{\lambda_1, \dots, \lambda_k\}}, \tag{8.39}$$

where $\lambda_1 \dots, \lambda_k$ *are the positive eigenvalues of* $F + F^\top$*, defined in* (8.29)*.*

(iii) If $F = \hat{F}$ *(the open-loop system is perfectly known) and assumption* ***C11*** *is satisfied, then the equilibrium point* (8.38) *is asymptotically stable for any* $\alpha > 0$*.*

Before proceeding, we make the following qualitative observation about the tuning parameters α and β, which we recall should satisfy (8.39). As the magnitudes of the eigenvalues of $F + F^\top$ increases, or the modeling error matrix $\tilde{F}$ decreases, the required value of the ratio $\frac{\alpha}{\beta}$ decreases. In other words, as the magnitude of damping present in the open-loop system increases or the uncertainty on the system decrease, less control action is required to ensure that the closed-loop system is stable.

8.5 Robust Integral Action for Fully Actuated Mechanical Systems

In Section 8.2, we show that the reject of constant disturbances in pH systems may be ensured adding an integral action on the passive output of pH systems. However, this methodology fails when regulation of the passive outputs of fully actuated mechanical systems is addressed. To verify this fact, we consider n-DOF, fully actuated mechanical system described in pH form by

$$\begin{bmatrix} \dot{q} \\ \dot{p} \end{bmatrix} = \begin{bmatrix} 0 & I_n \\ -I_n & -K_p \end{bmatrix} \nabla H(q,p) + \begin{bmatrix} 0 \\ I_n \end{bmatrix} u + \begin{bmatrix} d_u \\ d_a \end{bmatrix} \tag{8.40}$$

with Hamiltonian function

$$H(q,p) = \frac{1}{2} p^\top M^{-1}(q) p + V(q), \tag{8.41}$$

$q(t), p(t) \in \mathbb{R}^n$ are generalized positions and momenta, respectively, and are assumed measurable, $u \in \mathbb{R}^n$ is the control input, d_a and $d_u \in \mathbb{R}^n$ are the matched and unmatched disturbances – possibly time – varying, but bounded and unmeasurable. The inertia matrix $M(q) = M^\top(q) > 0$, and $K_p = K_p^\top > 0$ is the dissipation matrix. We assume that the Hamiltonian (8.41) has a minimum at the desired equilibrium $(q^\star, 0)$, that is, $q^\star = \arg\min V(q)$, and it is isolated. Note that $(q^\star, 0)$ is an asymptotically stable equilibrium of the mechanical system when $d_u = 0$ and $d_a = 0$ – the stability is almost global if $V(q)$ is proper and has a unique minimum (van der Schaft, 2016).

Thus, the standard addition of an integral action on the passive output, i.e. the velocities $\dot{q} = M^{-1}(q)p$ via

$$\begin{aligned} \dot{x}_c &= \dot{q} := M^{-1}(q)p \\ u &= -K_i x_c \end{aligned}$$

with $K_i = K_i^\top > 0$ has the following drawbacks. If d_u is a nonzero constant, the system admits no constant equilibrium, and if $d_u = 0$ and d_a is constant there is an equilibrium set given by

$$\mathcal{E} = \left\{ (q, p, x_c) \mid p = 0,\ \nabla V(q) + x_c = d_a \right\}.$$

Moreover, it is easy to see that, if $d_u = 0$ and d_a is constant, the foliation

$$\mathcal{M}_\kappa = \left\{ (q, p, x_c) \mid K_i q - x_c = \kappa,\ \kappa \in \mathbb{R}^n \right\},$$

is invariant with respect to the flow of the closed-loop system. Consequently, convergence to the desired equilibrium $(q^\star, 0, d_a)$ is attained only for a zero measure set of initial conditions. See Figure 8.1 for a pictorial description of the state space.

Indeed, from Proposition 8.1, we have boundedness of trajectories and stability of the equilibrium, however, the detectability requirement (iii) fails.

From the discussion above, it is clear that a more complex approach is required to reject the disturbances in mechanical systems. In this section, a PID-PBC controller based on the dynamic extension and change of coordinates described in Section 8.1.2 is proposed to tackle the problem.

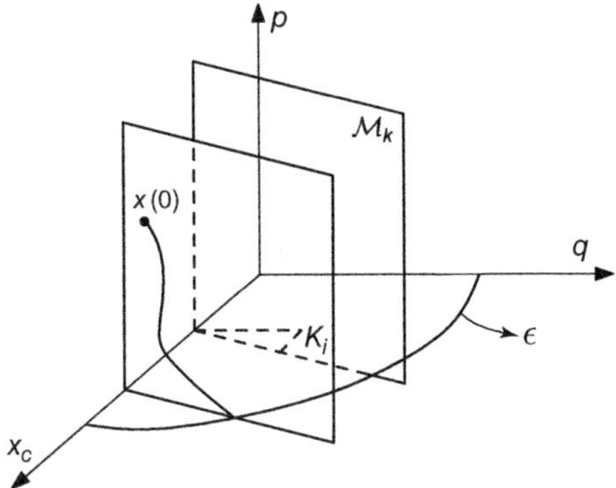

Figure 8.1 Graph of the state space showing two sheets of the invariant foliation $\mathcal{M}_\kappa$, the equilibrium set $\mathcal{E}$, and a trajectory $x(t) := ((q(t), p(t), x_c(t))$.

First, we show as a simple PI controller around the potential energy function with constant inertia matrix rejected both, matched and unmatched constant disturbance. Immediately, robust PID controllers are proposed for time-varying bounded disturbances for both, constant and nonconstant inertia matrix. For these controllers, the closed loop is endowing of the ISS property.

Lemma 8.1: *Consider the system* (8.40) *with constant inertia matrix M and* constant *disturbances* (d_u, d_a) *in closed-loop with the PI control*

$$\begin{aligned} \dot{x}_c &= K_i \nabla V(q) \\ u &= -K_p x_c - M K_i \nabla V(q), \end{aligned} \tag{8.42}$$

with $K_i = K_i^\top > 0$.

(i) *The closed-loop dynamics expressed in the coordinates,*

$$\begin{aligned} z_1 &= q \\ z_2 &= p + M(x_c - K_p^{-1} d_a) \\ z_3 &= x_c, \end{aligned} \tag{8.43}$$

takes the pH form

$$\dot{z} = \begin{bmatrix} 0 & I_n & -K_i \\ -I_n & -K_p & 0 \\ K_i & 0 & 0 \end{bmatrix} \nabla H_z(z), \tag{8.44}$$

with energy function

$$H_z(z) := H(z_1, z_2) + \frac{1}{2} || z_3 - z_3^\star ||_{K_i^{-1}}, \tag{8.45}$$

where $z_3^\star := d_u + K_p^{-1} d_a$.

(ii) The desired equilibrium point $z^\star := (q^\star, 0, z_3^\star)$, is asymptotically stable. The stability is almost global if $V(z_1)$ is proper and has a unique minimum.

One of the main, and rather intriguing, ideas of (Donaire and Junco, 2009) is the way the closed-loop energy function is constructed. Indeed, the first right-hand side of (8.45) is given by

$$H(z_1, z_2) = \frac{1}{2} z_2^\top M^{-1}(z_1) z_2 + V(z_1),$$

that is the *evaluation* of the function H given in (8.41) with the replacement $(q, p) \leftarrow (z_1, z_2)$. That is, of course, different from the composition of the function H with the change of coordinates defined in (8.43).

For time-varying disturbances, we can establish an *input-to-state stability* (ISS) property. Toward this end, the inclusion of a derivative term in ∇V_d is needed – in this case, with a constant derivative gain.[4]

Lemma 8.2: *Consider the system* (8.40) *with constant mass matrix M and* time-varying *disturbances $d(t) := \mathrm{col}(d_u(t), d_a(t))$, in closed-loop with the PID control law*

$$\begin{aligned} \dot{x}_c &= (M^{-1} + k_D R_3)\nabla V(q) + R_3 p \\ u &= -\left(K_3 R_3 + k_D \nabla^2 V(q) M^{-1}\right) p - K_4 \nabla V(q) - K_5 x_c, \end{aligned} \tag{8.46}$$

where k_D is a positive constant, $K_3 > 0$ and $R_3 > 0$ and

$$\begin{aligned} K_4 &:= k_D K_P M^{-1} + k_D K_3 R_3 + K_3 M^{-1} \\ K_5 &:= \left(K_P M^{-1} + M R_3\right) K_3. \end{aligned}$$

(i) The closed-loop dynamics expressed in the coordinates $z = \mathrm{col}(z_1, z_2, z_3)$ with

$$\begin{aligned} z_1 &= q \\ z_2 &= p + k_1 \nabla V(q) + K_3 x_c \\ z_3 &= x_c, \end{aligned}$$

takes the perturbed pH form

$$\dot{z} = \begin{bmatrix} -k_1 M^{-1} & I_n & -M^{-1} \\ -I_n & -K_P & -M R_3 \\ M^{-1} & R_3 M & -R_3 \end{bmatrix} \nabla H_z + \begin{bmatrix} I_n & 0 \\ k_1 \nabla^2 V(z_1) & I_n \\ 0 & 0 \end{bmatrix} d(t), \tag{8.47}$$

with new Hamiltonian[5] $H_z(z) = H(z_1, z_2) + \frac{1}{2}\|z_3\|_{K_3}$.

---

4 Recall that $\frac{d\nabla V_d}{dt} = \nabla^2 V_d M^{-1} p$.

5 To avoid cluttering, we use the same symbol to denote the energy function in all cases.

(ii) *If the potential energy function V is strictly convex with bounded Hessian, then* (8.47) *is* ISS *with respect to the time-varying input disturbances* $(d_1(t), d_2(t))$ *with ISS Lyapunov function* $H_z(z)$.

(iii) *If* $d_u = 0$ *and* d_a *is constant, then the desired equilibrium* $z^\star := (q^\star, 0, K_5^{-1} d_a)$ *is* asymptotically stable.

Remark 8.8: Comparing the closed-loop dynamics (8.44) and (8.47), we observe that, to enforce the ISS property, it was necessary to add damping in the $(1, 1)$ and $(3, 3)$ terms of the damping matrix of (8.47). This is achieved with the terms, added to the basic PI control (8.42), in (8.46).

The derivation of the controller for nonconstant inertia matrix M follows the same procedure used above. However, the PBC contains nonlinear PIDs and the expressions for the controller gains become quite involved, as shown below.

Lemma 8.3: *Consider the system* (8.40) *under the action of unmatched and matched time-varying disturbances* $d_u(t)$ *and* $d_a(t)$*, in closed-loop with the control law*

$$
\begin{aligned}
\dot{z}_3 &= \left[M^{-1} + k_1 R_3\right] \nabla V(q) + R_3 p \\
u &= -\left[k_1 K_P M^{-1} + K_3(M^{-1} + k_1 R_3)\right] \nabla V(q) - \left[k_1 \nabla^2 V(q) M^{-1} + K_3 R_3\right] p \\
&\quad - \left[\frac{1}{2}\sum_{i=1}^{n} e_i p^\top \nabla_{q_i} + K_P + F_{23}^\top M + \frac{1}{k_1}\left[I_n + F_{12}^\top\right] M F_{12}\right] \\
&\quad \times M^{-1} K_3 x_c + v(q, p)
\end{aligned}
\tag{8.48}
$$

where k_1 *is a positive constant,* $K_3 > 0$*, the mappings* $F_{12}, F_{23} : \mathbb{R}^n \times \mathbb{R}^n \times \mathbb{R}^n \to \mathbb{R}^{n\times n}$ *given by*

$$
\begin{aligned}
F_{12}(q,p,x_c) &:= -\frac{k_1}{2} M^{-1} \sum_{i=1}^{n} e_i \left[p + k_1 \nabla V + K_3 x_c\right]^\top M^{-1} \nabla_{q_i} M - I_n \\
F_{23}(q,p,x_c) &:= -\frac{1}{k_1} F_{12} + R_3 M
\end{aligned}
$$

and the mapping $v : \mathbb{R}^n \times \mathbb{R}^n \times \mathbb{R}^n \to \mathbb{R}^n$ *given as*

$$
\begin{aligned}
v(q,p,x_c) &:= \frac{k_1}{2} \sum_{i=1}^{n} e_i p^\top M^{-1} \nabla_{qi} M M^{-1} \nabla V - \left(M F_{12} M^{-1} + F_{12}^\top\right) \nabla V \\
&\quad - \frac{1}{k_1} F_{12}^\top M F_{12} M^{-1} \left[p + k_1 \nabla V\right].
\end{aligned}
$$

(i) *The closed-loop dynamics expressed in the coordinates* $z = \text{col}(z_1, z_2, z_3)$ *with*

$$z_1 = q$$
$$z_2 = p + k_1 \nabla V(q) + K_3 x_c$$
$$z_3 = x_c,$$

takes the perturbed pH form

$$\dot{z} = \begin{bmatrix} -k_1 M^{-1} & F_{12} & -M^{-1} \\ -F_{12}^\top & -K_P & -F_{23}^\top \\ M^{-\top} & F_{23} & -R_3 \end{bmatrix} \nabla H_z + \begin{bmatrix} I_n & 0 \\ k_1 \nabla^2 V(z_1) & I_n \\ 0 & 0 \end{bmatrix} \begin{bmatrix} d_u(t) \\ d_a(t) \end{bmatrix} \tag{8.49}$$

with $H_z(z) = H(z_1, z_2) + \frac{1}{2}\|z_3\|_{K_3}$.

(ii) *The closed-loop system is ISS with respect to the disturbances* $(d_u(t), d_a(t))$, *provided that the Hessian of the potential energy satisfies condition (ii) in Lemma 8.1.*

(iii) *The* unperturbed *system*(8.49) *has an asymptotically stable equilibrium at the desired state* $z^\star = (q^\star, 0, 0)$.

Remark 8.9: Comparing Lemmas 8.2 and 8.3, we observe that if the damping term R_3 is removed, only the weaker property of Integral Input-to-State Stability (IISS) with respect to matched disturbances, can be established. This result, for constant M, is omitted for brevity.

Remark 8.10: Making use of the transformed mechanical system proposed in Venkatraman et al. (2010) and Romero et al. (2013) other variations of these PID controllers, that yield simpler expressions for some particular cases are presented. The interested reader is referred to the latter paper for further details.

8.6 Robust Integral Action for Underactuated Mechanical Systems

IDA-PBC, first introduced in Ortega et al. (2001), is a highly popular controller design technique applicable for equilibrium stabilization of a wide class of physical systems (see e.g. (Aoki et al., 2016; Cai et al., 2007; Shimizu et al., 2009)). A comprehensive discussion of IDA-PBC may be found in Ortega and García-Canseco (2004). Its application for *underactuated mechanical systems* has been particularly successful as reported, for instance, in Acosta et al. (2005), Mahindrakar et al. (2006), Ortega et al. (2002), and Viola et al. (2007). It is widely recognized that IDA-PBC designs are robust against parameter uncertainties and unmodeled dynamics, e.g., passive effects like friction. However, the (unavoidable) presence of external disturbances degrades its performance, shifting the equilibrium of the closed-loop and, possibly, inducing instability. For this reason, the problem of robustification of IDA-PBC *vis-à-vis* external disturbances is of primary importance.

For *fully actuated* mechanical systems, this problem has been addressed in Section 8.3, where the key idea of adding an integral action that preserves the pH structure of the system, proposed in Section 8.1, is exploited. However, extending this robustification result to the case of underactuated mechanical systems is a very challenging problem.

In this section, we propose an outer-loop controller that solves the problem of (constant, matched) disturbance rejection for underactuated n-degress of freedom mechanical systems with arbitrary underactuation degree. The controller is designed adding an integral action on nonpassive outputs of underactuated systems, with the developments of Section 8.3. An interesting feature of the proposed outer-loops is that they do not destroy the mechanical structure of the system, preserving in closed-loop its pH form.

The design is applicable for systems where the mass matrix is *independent of the unactuated* coordinates and the closed-loop mass matrix is *constant*. The first assumption is instrumental for the results reported in Acosta et al. (2005), Mahindrakar et al. (2006), and Viola et al. (2007), where it is imposed to simplify the kinetic energy matching equation. In the present work this assumption, as well as the requirement that the closed-loop mass matrix is constant, is needed to construct a suitable change of coordinates under which the integral action is added. It should be noted that these assumptions are verified by a large class of underactuaded mechanical systems, including those considered in Mahindrakar et al. (2006).

8.6.1 Standard Interconnection and Damping Assignment PBC

IDA-PBC was introduced in Ortega et al. (2002) to control underactuated mechanical systems described in pH form (D.3) with the matrix $G(q)$ verifying $\text{rank}(G) = m < n$.

The control objective is to design a static, state-feedback that assigns to the closed-loop a desired stable equilibrium $(q, p) = (q^\star, 0)$, $q^\star \in \mathbb{R}^n$. This is achieved in IDA-PBC by matching the pH target dynamics

$$\begin{bmatrix} \dot{q} \\ \dot{p} \end{bmatrix} = \begin{bmatrix} 0 & M^{-1}(q)\, M_d(q) \\ -M_d(q)\, M^{-1}(q) & J_2(q,p) - R_d(q) \end{bmatrix} \nabla H_d \tag{8.50}$$

with the new total energy function $H_d : \mathbb{R}^n \times \mathbb{R}^n \to \mathbb{R}$,

$$H_d(q,p) := \frac{1}{2} p^\top M_d^{-1}(q)\, p + V_d(q), \tag{8.51}$$

where the desired mass matrix $M_d : \mathbb{R}^n \to \mathbb{R}^{n\times n}$ is positive definite, the desired potential energy $V_d : \mathbb{R}^n \to \mathbb{R}$ verifies

$$q^\star = \arg\min \;\; V_d(q), \tag{8.52}$$

and the desired damping matrix is defined by

$$R_d(q) := G(q)K_P G^\top(q) \geq 0,$$

with $K_P \in \mathbb{R}^{m\times m}$ a free positive definite matrix. The matrix $J_2 : \mathbb{R}^n \times \mathbb{R}^n \to \mathbb{R}^{n\times n}$ is free to the designer and fulfills the skew-symmetry condition

$$J_2(q,p) = -J_2^\top(q,p). \tag{8.53}$$

The closed-loop system (8.50) has a stable equilibrium point at $(q^\star, 0)$ with Lyapunov function H_d, which verifies

$$\dot{H}_d = -\|G^\top M_d^{-1} p\|^2_{K_P} \leq 0. \tag{8.54}$$

The closed-loop is asymptotically stable provided that the output

$$y_d := G^\top M_d^{-1} p \tag{8.55}$$

is detectable (van der Schaft, 2016).

By equating the right-hand sides of (D.3) and (8.50), one obtains the so-called matching equations, which are two PDEs that identify the assignable M_d and V_d, and gives an explicit expression for the (static state-feedback) control signal $u = u_{\mathrm{IDA}}(q,p)$, where $u_{\mathrm{IDA}} : \mathbb{R}^n \times \mathbb{R}^n \to \mathbb{R}^m$.

Formulation of the Robust IDA-PBC Problem

We consider the effect of constant, matched disturbances in the mechanical system (D.3) that may represent external forces or an input measurement bias. We assume that the system (D.3) is in closed loop with an IDA-PBC into which the disturbance propagates and must be rejected with a *dynamic outer-loop control*. That is, we consider the system (D.3) perturbed by an input disturbance in closed-loop with the control $u = u_{\mathrm{IDA}}(q,p) + v$, leading to the following.

Problem formulation. Given the dynamics

$$\begin{bmatrix} \dot{q} \\ \dot{p} \end{bmatrix} = \begin{bmatrix} 0 & M^{-1} M_d \\ -M_d M^{-1} & J_2 - R_d \end{bmatrix} \nabla H_d + \begin{bmatrix} 0 \\ G \end{bmatrix} (v + d_a), \tag{8.56}$$

with H_d as in (8.51), $d_a \in \mathbb{R}^m$, and $R_d = GK_pG^\top$. Find a dynamic controller $u(q, p, x_c)$, where $x_c \in \mathbb{R}^m$ is the state of the controller, that ensures asymptotic stability of the desired equilibrium $(q, p, x_c) = (q^\star, 0, x_c^\star)$, for some $x_c^\star \in \mathbb{R}^m$, even under the action of the constant disturbance d_a.

Remark 8.11: Notice that the dimension of the dynamic extension x_c coincides with the one of the input space, i.e. m. As will be shown below, this choice suffices to provide a solution to the problem.

8.6.2 Main Result

To present the main result of this section, we impose the following assumption which characterizes the class of mechanical systems that may be addressed:

C12 The input matrix G and the desired mass matrix M_d are constant, and the mass matrix $M(q)$ is independent of the nonactuated coordinates. Consequently,

$$G^\perp \nabla_q (p^\top M^{-1} p) = 0,$$

where $G^\perp \in \mathbb{R}^{(n-m)\times n}$ is a full-rank left annihilator of G.

We have that the term $G^\perp \nabla_q (p^\top M^{-1} p)$ appears in the kinetic energy matching equation as a forcing term that makes the PDE inhomogeneous and introduces a quadratic term in the unknown M_d, rendering very difficult its solution (Ortega et al., 2002). In Acosta et al. (2005), it is also assumed to be zero to provide an explicit solution to the PDE. In Proposition 2 of (Viola et al., 2007), it is shown that a sufficient condition to eliminate this term is that the Coriolis and centrifugal forces of the mechanical system enter into the kernel of G.

Now, we are in position to announce the main result and it is described in the following Proposition:

Proposition 8.9: *Consider the dynamics* (8.56), *verifying assumption* ***C12*** *in closed-loop with the PID controller*

$$\dot{x}_c = \left(K_2^\top G^\top M^{-1} + K_3^\top G^\top M_d^{-1} G K_1 G^\top M^{-1}\right) \nabla V_d + K_3^\top G^\top M_d^{-1} p, \quad (8.57)$$

$$u = -\left[K_p G^\top M_d^{-1} G K_1 G^\top M^{-1} + K_1 G^\top \frac{dM^{-1}}{dt} + K_2 K_I \right.$$
$$\left. \times \left(K_2^\top + K_3^\top G^\top M_d^{-1} G K_1\right) G^\top M^{-1} \frac{1^A}{2}\right] \nabla V_d$$

$$
\begin{aligned}
&- \left[\frac{1}{2}K_1 G^\top M^{-1}\nabla^2 V_d M^{-1} + (G^\top G)^{-1}G^\top J_2 M_d^{-1} + K_2 K_I K_3^\top G^\top M_d^{-1}\right] p \\
&- \left(K_P + K_3\right) K_I x_c
\end{aligned}
\tag{8.58}
$$

where $K_1 > 0$, $K_P > 0$, *and* $K_I > 0$, $K_3 > 0$ *and* $K_2 := (G^\top M_d^{-1} G)^{-1}$.

(i) *The closed-loop dynamics in the coordinates* $z = \mathrm{col}(z_1, z_2, z_3)$ *with*

$$
\begin{aligned}
z_1 &= q \\
z_2 &= p + GK_1 G^\top M^{-1}\nabla V_d(q) + GK_2 K_I (x_c - z_3^\star) \\
z_3 &= x_c,
\end{aligned}
\tag{8.59}
$$

with $z_3^\star := K_I^{-1}(K_P + K_3)^{-1} d_a$, *can be written in pH form as follows:*

$$
\begin{bmatrix} \dot z_1 \\ \dot z_2 \\ \dot z_3 \end{bmatrix} = \begin{bmatrix} -R_{11} & M^{-1} M_d & -F_{13} \\ -M_d M^{-1} & -GK_P G^\top & -GK_3 \\ F_{13}^\top & K_3^\top G^\top & -R_{33} \end{bmatrix} \nabla H_z
\tag{8.60}
$$

with Hamiltonian

$$
H_z(z) = \frac{1}{2} z_2^\top M_d^{-1} z_2 + V_d(z_1) + \frac{1}{2}\|z_3 - z_3^\star\|_{K_I}^2,
$$

and the mappings $R_{11} : \mathbb{R}^n \to \mathbb{R}^{n\times n}$, $F_{13} : \mathbb{R}^n \to \mathbb{R}^{n\times m}$, $R_{33} : \mathbb{R}^n \to \mathbb{R}^{m\times n}$ *given by*

$$
\begin{aligned}
R_{11}(q) &:= M^{-1} GK_1 G^\top M^{-1}, \quad F_{13}(q) := M^{-1} GK_2, \\
R_{33}(q) &= K_3^\top G^\top M_d^{-1} GK_2.
\end{aligned}
$$

(ii) *The equilibrium* $(q, p, x_c) = (q^\star, 0, z_3^\star)$ *is* stable.

(iii) *If the output*

$$
y_{D3} = \begin{bmatrix} G^\top M^{-1}\nabla V_d \\ G^\top M_d^{-1} z_2 \\ K_I (z_3 - z_3^\star) \end{bmatrix}
$$

is a detectable output of the dynamics (8.60), *then* $(q^\star, 0, z_3^\star)$ *is an* asymptotically *stable equilibrium of the closed-loop.*

In Donaire et al. (2017) two additional controllers, which are simplified versions of the one given in Proposition 8.7, are presented. These two controllers are obtained setting $(K_1, K_3) = (0,\ 0)$ and $(K_2, K_3) = (0, I_m)$. As seen in (8.60), these modifications still preserve the pH structure but eliminate some damping terms. Consequently, their corresponding detectability condition is strictly stronger than (iii) above, reducing the class of systems for which asymptotic stability is guaranteed. See Donaire et al. (2017) for further details.

8.7 A New Robust Integral Action for Underactuated Mechanical Systems

In Section 8.6 a robust controller to reject matched disturbances in underactuated mechanical systems is designed making use of a change of coordinates and the strong assumption **C12**, which limits the applicability to a class of system. In this section, a new control design is presented to address the same problem but where both conditions are conspicuous by their absence.

8.7.1 System Model

In this section, the IA methodology proposed in Section 8.4.2 is exploited to robustify energy-shaping controlled underactuated mechanical systems with respect to constant matched disturbances. The considered class of system has the dynamics (8.56) with frictions characterized by the matrix $R_d(q)$.

In practice, it is commonplace for the model of the open-loop mechanical system to contain uncertainty regarding the dissipative (friction) terms. When applying IDA-PBC, this form of uncertainty can be accounted for when applying the technique of "*passivation by damping injection*"' – see Section 3.2 of (Gómez-Estern and van der Schaft, 2004). Application of this technique results in a dissipation matrix $R_d(q) \in \mathbb{R}^{n\times n}$ that is not symmetric, but rather, has the form

$$R_d(q) = D_p(q)M^{-1}(q)M_d(q) + G_u(q)K_dG_u^\top(q), \tag{8.61}$$

where $D_p(q) = D_p^\top(q) \in \mathbb{R}^{n\times n}$ is the positive definite damping matrix of the open-loop system and $K_d = K_d^\top \in \mathbb{R}^{n\times n}$ is a positive definite tuning parameter. Importantly, the closed-loop damping matrix (8.61) satisfies $R_d + R_d^\top > 0$.

IDA-PBC ensures that V_d has an isolated minimum at a desired value $q^\star$. The latter fact, together with positivity of the matrix M_d, implies that $(q^\star, 0)$ is a stable equilibrium of (8.56) – with Lyapunov function H_d – when u and d_a are zero. The robust regulation control objective is to design an integral action control that ensures stability of the desired equilibrium $(q, p, x_c) = (q^\star, 0, x_c^\star)$ for some $x_c \in \mathbb{R}^m$.

8.7.2 Coordinate Transformation

Application of the IA methodology of Section 8.4.2 to the mechanical system (8.56) is stymied by the fact that the input mapping matrix is not of the form (8.4). This issue is addressed by recalling Lemma 1 of (Ferguson et al., 2017a), which showed that the system (8.56) expressed in the coordinates

$$x = f_p(x_0) := \begin{bmatrix} Q(q)p \\ q \end{bmatrix}, \tag{8.62}$$

where $x_0 := \text{col}(q, p)$,

$$Q(q) := \begin{bmatrix} (G^\top(q)G(q))^{-1}G^\top(q) \\ G^\perp(q) \end{bmatrix} \tag{8.63}$$

and $G^\perp(q) \in \mathbb{R}^{(n-m)\times n}$ is a left annihilator of the matrix G, has an input mapping matrix of the form (8.4). Letting the Jacobian of f_p be denoted by

$$K(x) := \nabla^\top f_p(q,p) = \begin{bmatrix} \nabla_q^\top [Q(q)p] & Q(q) \\ I_n & 0 \end{bmatrix}, \tag{8.64}$$

the energy-shaping controlled mechanical system (8.56) can be expressed in the form (8.1) with

$$\begin{aligned} F(x) &= K(x_0)F_0(x_0)K^\top(x_0)|_{x_0 = f_p^{-1}(x)}, \\ F_0 &= \begin{bmatrix} 0 & M^{-1}M_d \\ -M_d M^{-1} & J_2 - R_d \end{bmatrix}, \\ \nabla H(x) &= K^{-\top}(x_0)\nabla H_d(x_0)|_{x_0 = f_p^{-1}(x)}, \\ G &= K(x_0)\begin{bmatrix} 0 & G^\top(q) \end{bmatrix}^\top = \begin{bmatrix} I_m & 0 \end{bmatrix}^\top. \end{aligned} \tag{8.65}$$

8.7.3 Verification of Requisites

Using the system matrix definitions (8.65), the IA (8.35) of Proposition 8.7 can now be tailored to the mechanical system. Before application, however, assumption **C10** must be verified. To this end, recall that it was assumed that the only modeling error for the mechanical system is associated with the term R_d. This means that the estimated F_0 matrix has the form

$$\hat{F}_0(x_0) = \begin{bmatrix} 0 & M^{-1}(q)M_d(q) \\ -M_d(q)M^{-1}(q) & J_2(q,p) - \hat{R}_d(q) \end{bmatrix}. \tag{8.66}$$

Unwinding the definitions of $F(x)$ in (8.65), the modeling error matrix $\tilde{F}(x)$ is given by

$$\tilde{F}(x) = K\tilde{F}_0K^\top = \begin{bmatrix} -Q\tilde{R}_dQ^\top & 0 \\ 0 & 0 \end{bmatrix}, \tag{8.67}$$

where $\tilde{F}_0(x) := F_0(x) - \hat{F}_0(x)$ and $\tilde{R}_d(x) := R_d(x) - \hat{R}_d(x)$. By similar argument, the damping matrix R is given by

$$R = K\text{sym}(F_0)K^\top = \begin{bmatrix} Q\text{sym}(R_d)Q^\top & 0 \\ 0 & 0 \end{bmatrix}. \tag{8.68}$$

As $\mathrm{sym}(R_d) > 0$, the kernel of R is spanned by the columns of

$$U_u = \begin{bmatrix} 0 \\ I_n \end{bmatrix}. \tag{8.69}$$

Clearly, the modeling error matrix satisfies $\tilde{F}U_u = 0$, verifying assumption **C10**(a). Assumption **C10**(b) is satisfied precisely when $\|\tilde{F}\| = \|Q\tilde{R}_dQ^\top\| < \kappa$ for some $\kappa \geq 0$. That is, the error in the open-loop friction model should be bounded in the domain of interest.

8.7.4 Robust Integral Action Controller

As assumption **C10** has been verified, the control law (8.35) can be applied to the system (8.56), with damping (8.61), to reject the effects of the constant disturbance d_a by utilizing the definitions (8.65). The control signal can now be evaluated as

$$u = -\alpha K_i \left[\beta(G^\top G)^{-1}G^\top p - x_c\right]. \tag{8.70}$$

Likewise, the update law can be evaluated

$$\begin{aligned} \dot{x}_c = {} & \beta\nabla_q^\top \left[(G^\top G)^{-1}G^\top p\right] M^{-1}M_d\nabla_p H_d \\ & -\beta(G^\top G)^{-1}G^\top M_d M^{-1}\nabla_q H_d \\ & + \beta(G^\top G)^{-1}G^\top \left[J_2 - \hat{R}_d\right] \nabla_p H_d \\ & - \alpha G^\top \nabla_p H_d. \end{aligned} \tag{8.71}$$

Recalling the condition (8.39), the IA (8.70), (8.71) is stable for any α and β satisfying

$$\frac{\alpha}{\beta} > \frac{\kappa^2}{4\lambda_{\min}\left[Q\mathrm{sym}(R_d)Q^\top\right]}. \tag{8.72}$$

8.8 Examples

In this section, we illustrate the behavior of the robust controllers formulated in the Sections 8.5, 8.6, and 8.7. First, we show how the IA proposed in Lemma 8.1 can be used to compensate for bias in the velocity measurement. Then, we present simulations of the controller proposed in Lemma 8.3 using a prismatic robot. To validate the controller of Proposition 8.9, we present simulations for the Acrobot system and immediately experimental results using the disk on disk system. Finally, the damped vertical take-off and landing (VTOL) aircraft example is used to illustrate the applicability of the controller described in Section 8.7.

8.8.1 Mechanical Systems with Constant Inertia Matrix

In this section, we consider an n-DOF, fully actuated, fully damped, perturbed mechanical system represented in pH form (8.40), with a constant disturbance d_u and $d_a = 0$. The energy function if given by

$$H(q,p) = \frac{1}{2}p^\top M^{-1}p + V(q), \tag{8.73}$$

with $M \in \mathbb{R}^{n\times n}$ the positive definite, constant inertia matrix, and $V(q)$ is the potential energy function. We assume that $q^\star = \arg\min V(q)$, and it is isolated and global. Then, by Lemma 8.1 the IA

$$\begin{aligned} \dot{x}_c &= K_i \nabla V \\ u &= -K_p x_c - MK_i \nabla V, \end{aligned} \tag{8.74}$$

with $K_i = K_i^\top > 0$, ensures the equilibrium $(q^\star, -Md_u, d_u)$ is GAS with Lyapunov function

$$W = \frac{1}{2}(p + Mx_c)^\top M^{-1}(p + Mx_c) + V(q) + \frac{1}{2}(x_c - d_u)^\top K_i^{-1}(x_c - d_u).$$

The disturbance considered in the example represents a bias term in the measurement of velocity that propagates into the system through the damping injection. This fact is clear writing the dynamics of the open-loop system in Euler–Lagrange form

$$M\ddot{q} + K_p(\dot{q} - d_1) + \nabla V(q) = u.$$

It is interesting to note that, after differentiation, the closed-loop system is given by

$$M\dddot{q} + K_p\ddot{q} + (I_n + MK_i)\nabla^2 V(q)\dot{q} + K_pK_i\nabla V(q) = 0.$$

Hence, the stabilization mechanism is akin to the introduction of nonlinear *gyroscopic forces* plus a suitable weighting of the potential energy term.

8.8.2 Prismatic Robot

In this section, we use the two DOF prismatic robot example of (Angeli et al., 2000) to illustrate in simulations of Lemma 8.3. Similar to Angeli et al. (2000), the initial condition vector is $[q_{1_0}, q_{2_0}, p_{1_0}, p_{2_0}, z_{31_0}, z_{32_0}] = [0, 0.1, 0.1, 0.1, 0.1, 0.2]$ and the desired equilibrium is the origin. The bounded disturbance vector is taken as $d_a = \alpha \tanh(\dot{q})$, with $\alpha = 3, 10$. The parameters of the model are the same as in Angeli et al. (2000) and are repeated here for ease of reference. The mass matrix is

$$M = \begin{bmatrix} m_1 q_2^2 + \frac{m_2 L^2}{3} & 0 \\ 0 & m_1 \end{bmatrix}$$

where m_1 and L is the mass and length of the arm and m_2 is the mass of the hand. The states $q = [q_1\ q_2]^\top$ and $p = [p_1\ p_2]^\top$ are the generalized position and momenta, respectively. The subscript 1 and 2 indicates variables of the arm and the hand, respectively. The system has no potential energy and no dissipation.

In Remark 8.9, we mentioned that if R_3 is removed of the controller (8.48), then the closed loop would be endowed of the IISS property. Hence, we present simulations with two controllers – denoted IISS and ISS Controllers. The motivation for the names stems from the fact that the first controller ensures only IISS, while the second one strengthens this to ISS. As will be illustrated below, although both controllers yield bounded trajectories, the transient behavior of the ISS controller is far superior. As is well known, IISS does not ensure bounded-input-bounded-state behavior – for all inputs – but in this particular case, they turn out to be bounded. It should be remarked that in Angeli et al. (2000) it is claimed that the trajectories are unbounded. The problem is that, to observe this fact, the simulation has to run in a longer horizon than the one used in Angeli et al. (2000).

The expression for the IISS controller is

$$
\begin{aligned}
u &= -(k_1K_pM^{-1} + I_n)K_d\tilde{q} - (k_1+1)K_dM^{-1}p - z_3 \\
&\quad + \begin{bmatrix} \frac{3k_1k_{d2}q_2\tilde{q}_2}{3mq_2^2+ML^2}(p_1 + k_1k_{d1}\tilde{q}_1) - k_1mq_2^2(p_1 + k_1k_{d1}\tilde{q}_1)^3 \\ \frac{9k_1k_{d1}q_2\tilde{q}_1}{(3mq_2^2+ML^2)^2}[(1+m)p_1 + k_1k_{d1}m\tilde{q}_1] \end{bmatrix} \\
\dot{z}_3 &= K_iM^{-1}\left[p + k_1K_d\tilde{q}\right],
\end{aligned}
\tag{8.75}
$$

with $\tilde{q} = q - q^\star$. The expression for the ISS controller is

$$
\begin{aligned}
u &= -(k_1K_pM^{-1} + I_n)K_d\tilde{q} - (k_1+1)K_dM^{-1}p - (K_pM^{-1} + MR_3)K_3z_3 \\
&\quad - \begin{bmatrix} -\frac{27k_1mq_2^2(p_1+k_1k_{d1}\tilde{q}_1+k_{31}z_{31})^2}{(3mq_2^2+ML^2)^3} & \frac{3q_2(p_1+k_1k_{d1}\tilde{q}_1+k_{31}z_{31})}{(3mq_2^2+ML^2)} \\ \frac{9mq_2(k_1k_{d1}\tilde{q}_1+k_{31}z_{31})}{(3mq_2^2+ML^2)^2} & 0 \end{bmatrix} K_3z_3 + \\
&\quad + \begin{bmatrix} \frac{3k_1k_{d2}q_2\tilde{q}_2}{3mq_2^2+ML^2}(p_1 + k_1k_{d1}\tilde{q}_1 + k_{31}z_{31}) - k_1mq_2^2(p_1 + k_1k_{d1}\tilde{q}_1 + k_{31}z_{31})^3 \\ \frac{9k_1k_{d1}q_2\tilde{q}_1}{(3mq_2^2+ML^2)^2}[(1+m)p_1 + k_1k_{d1}m\tilde{q}_1 + k_{31}z_{31}] \end{bmatrix} \\
\dot{z}_3 &= R_3p + \left[M^{-1} + k_1R_3\right]K_d\tilde{q}.
\end{aligned}
\tag{8.76}
$$

The values of the model and controllers parameters are as follows: $m_1 = 1$, $ML^2 = 3$, $K_p = \mathrm{diag}(2,1)$, $K_d = \mathrm{diag}(k_{d1}, k_{d2}) = \mathrm{diag}(2,1)$, $R_3 = \mathrm{diag}(4,4)$, $K_i = K_3 = \mathrm{diag}(k_{31}, k_{32}) = \mathrm{diag}(3,3)$, and $k_1 = 2$.

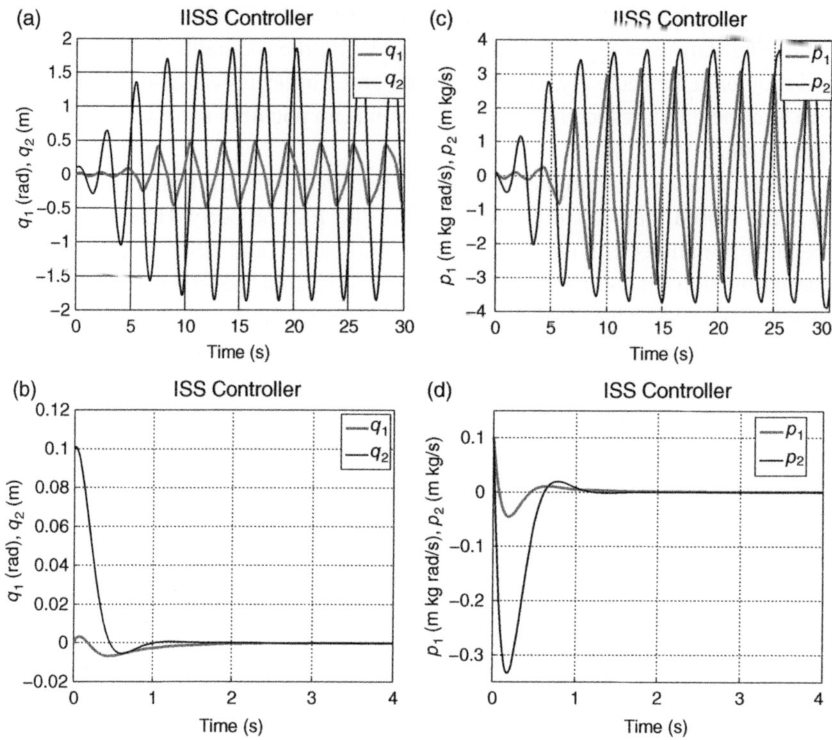

Figure 8.2 (a, b) Angle of the arm q_1 and position of the hand q_2, and (c, d) momenta of the arm p_1 and the hand p_2 for $\alpha = 3$.

Figures 8.2 and 8.3 show the behavior of the system for the smaller disturbance, that is $\alpha = 3$, while the case of $\alpha = 10$ is depicted in Figures 8.4 and 8.5. In all cases, the superior performance of the ISS controller is evident. It is interesting to note that the improved performance is not achieved injecting larger gains in the loop. Actually, as shown in the second column of Figures 8.3 and 8.5, which show the control signals, the control action of the IISS controller is far more demanding than that of the ISS controller.

The bounded disturbances acting on the system are shown in the third column of Figures 8.3 and 8.5. As expected, the performance is deteriorated for bigger disturbances. However, the ISS controller still shows acceptable transients. On the other hand, the behavior of the IISS controller might not be practically acceptable – as the demanded forces might exceed the actuator limits.

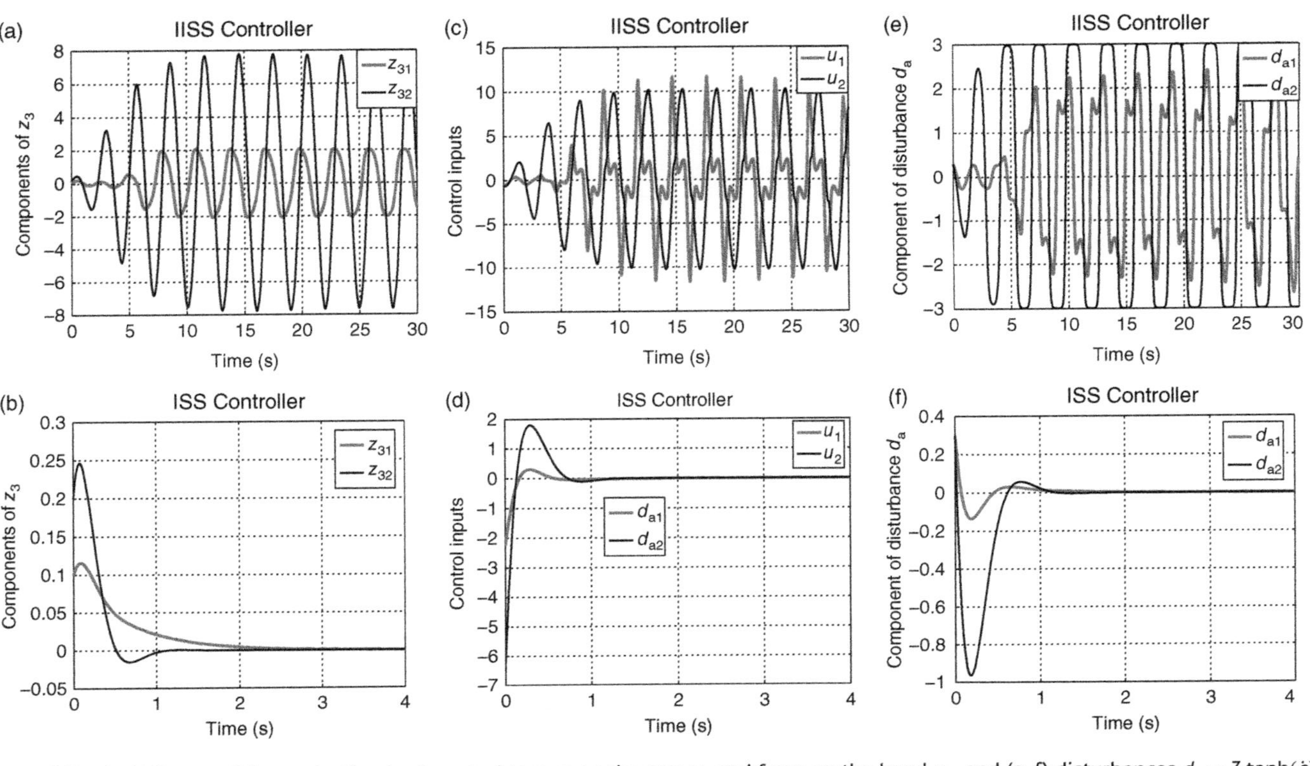

Figure 8.3 (a, b) States of the controller, (c, d) control torque on the arm u_1 and force on the hand u_2, and (e, f) disturbances $d_a = 3\tanh(\dot{q})$ for $\alpha = 3$.

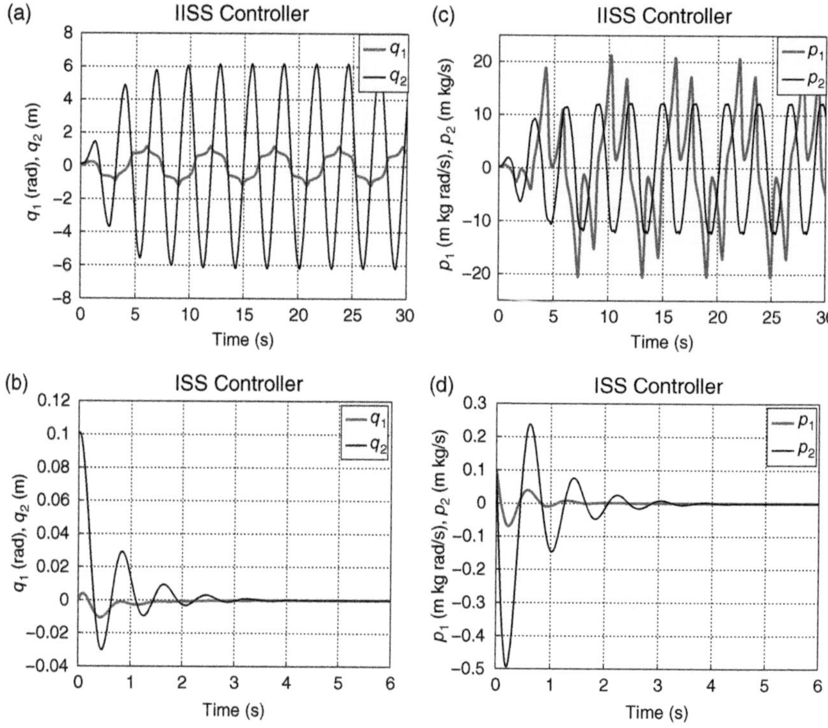

Figure 8.4 (a, b) Angle of the arm q_1 and position of the hand q_2, and (c, d) momenta of the arm p_1 and the hand p_2 for $\alpha = 10$.

8.8.3 The Acrobot System

The equations of motion of the system are given by (8.56) with $n = 1, m = 1$,

$$M(q_2) = \begin{bmatrix} c_1 + c_2 + 2c_3\cos(q_2) & c_2 + c_3\cos(q_2) \\ c_2 + c_3\cos(q_2) & c_2 \end{bmatrix},$$

$$V(q) = g\left[c_4\cos(q_1) + c_5\cos(q_1 + q_2)\right],$$

$$G = \begin{bmatrix} 0 \\ 1 \end{bmatrix},$$

where g is the gravitational constant and c_1, c_2, c_3, and c_4 are constant parameters of the system with $c_1 \neq c_2$. The upright equilibrium $q^\star = (0, 0)$ of the Acrobot can be stabilized asymptotically with the IDA-PBC controller

$$u_{\mathrm{IDA}}(q,p) = \frac{1}{2}\nabla_{q_2}(p^\top M^{-1}p) + \nabla_{q_2}V - \begin{bmatrix} k_2 & k_3 \end{bmatrix} M^{-1}\nabla V_d + \frac{k_v}{k_1k_3 - k_2^2}(k_2p_1 - k_1p_2), \tag{8.77}$$

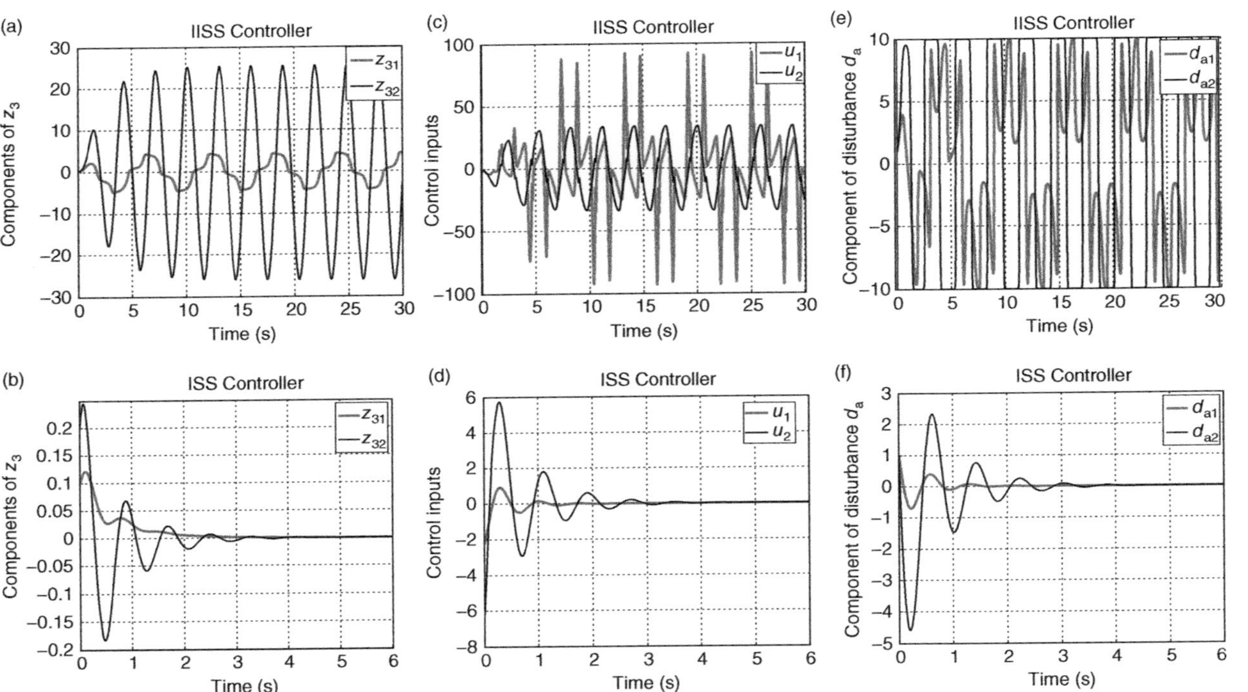

Figure 8.5 (a, b) States of the controller, (c, d) control torque on the arm u_1 and force on the hand u_2, and (e, f) disturbances $d_a = 3\tanh(\dot{q})$ for $\alpha = 10$.

where $k_v > 0$ is the damping injection gain, and the controller gains k_1, k_2, and k_3 with $k_1 := \left(1 - \sqrt{c_1/c_2}\right) k_2 > 0$, and $k_3 > \frac{k_2}{1-\sqrt{c_1/c_2}}$. The desired mass matrix is

$$M_d = \begin{bmatrix} k_1 & k_2 \\ k_2 & k_3 \end{bmatrix} > 0,$$

and the desired potential energy is such that

$$\begin{aligned}
\nabla_{q_1} V_d &= -k_0 \sin(q_1 - \mu q2) - b_1 \sin(q_1) - b_2 \sin(q_1 + q_2) \\
&\quad -b_3 \sin(q_1 + 2q_2) - b_4 \sin(q_1 - q_2) + k_u(q_1 - \mu q_2), \\
\nabla_{q_2} V_d &= k_0 \mu \sin(q_1 - \mu q2) - b_2 \sin(q_1 + q_2) - 2b_3 \sin(q_1 + 2q_2) \\
&\quad +b_4 \sin(q_1 - q_2) - k_u \mu (q_1 - \mu q_2),
\end{aligned}$$

with the constants

$$\begin{aligned}
b_1 &:= \frac{g}{2k_2}(c_3c_5 \pm 2\sqrt{c_1c_2}c_4), \; b_2 := \frac{g\mu}{2k_2(\mu+1)}(c_3c_4 \pm 2\sqrt{c_1c_2}c_5) \\
b_3 &:= \frac{g\mu c_3c_5}{2k_2(\mu+2)}, \quad b_4 := \frac{g\mu c_3c_4}{2k_2(\mu-1)}, \quad \mu := \frac{-1}{1+\sqrt{\frac{c_1}{c_2}}},
\end{aligned}$$

and arbitrary constant k_0.

For the simulations, we use the values of the model parameters provided in Mahindrakar et al. (2006), that is, $g = 9.8$, $c_1 = 2.3333$, $c_2 = 5.3333$, $c_3 = 2$, $c_4 = 3$, $c_5 = 2$. The simulations are performed under the following extreme scenario: the system starts with the Acrobot in closed loop with the IDA-PBC hanging down with zero velocity, that is with initial conditions $q_1(0) = -\pi$, $q_2(0) = 0$, $p_1(0) = 0$ and $p_2(0) = 0$ and a matched constant disturbance $d_a = 10$ N m is added to actuated link of the system at time $t = 25$s. The gains of the IDA-PBC (8.77) were selected as follows: $k_1 = 0.3386$, $k_2 = 1$, $k_3 = 5.9073$, $\mu = -0.6019$, $k_0 = -260$, $k_u = 60$, $k_v = 70$.

First, we observe that G and the desired mass matrix are constant and $G^{\perp}\nabla_q(p^{\top}M^{-1}p) = 0$, thus assumption **C12** is verified. To reject the disturbances, we add to the IDA-PBC the outer-loop controller (8.57), (8.58) of Proposition 8.9 with parameters $K_1 = 0.005$, $K_3 = 25$, $K_I = 0.02$, and $K_P = k_v$.

Figures 8.6 and 8.7 show the time histories of the Acrobot's angles and angular velocities. It is clear from the plots that the angles converge to the desired position, while the velocities converge to zero. Figure 8.8 shows the time history of the controller state x_c, which provides the disturbance rejection. Note that the plot of x_c in Figure 8.8 has been multiplied by the constant $\mathcal{K}_x = (K_P + K_3)K_I$, such that $\mathcal{K}_x x_c$ converges to $\mathcal{K}_x z_3^{\star} = d_a$.

A video animation of the Acrobot in closed loop with both IDA-PBC and IDA-PBC plus the PID controller can be watched on

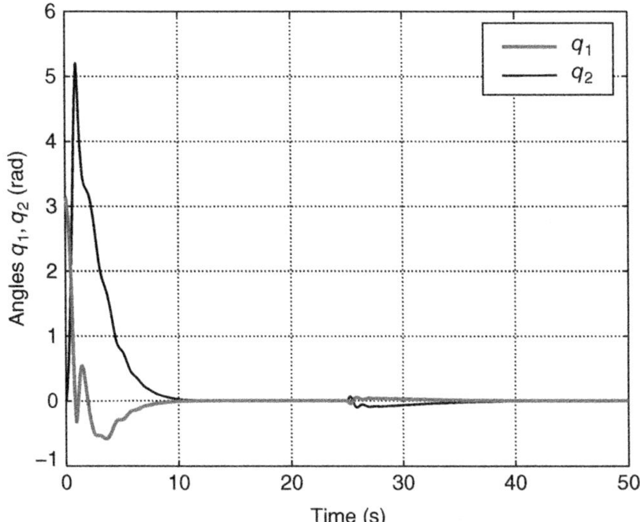

Figure 8.6 Time histories of the Acrobot angles q_1 and q_2 with the IDA-PBC plus the nonlinear PID.

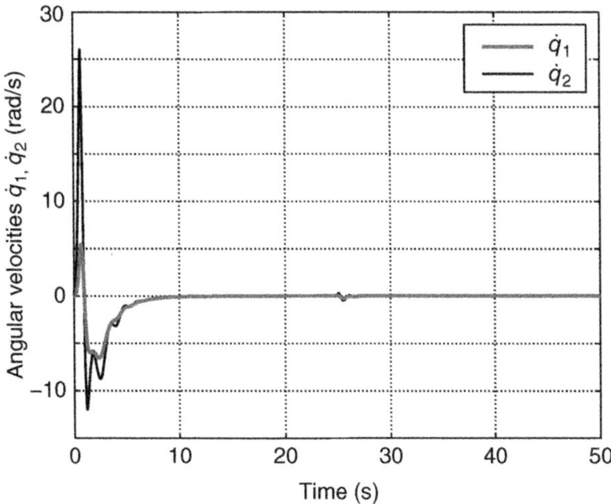

Figure 8.7 Time histories of the Acrobot angular velocities $\dot{q}_1$ and $\dot{q}_2$ with the IDA-PBC plus the nonlinear PID.

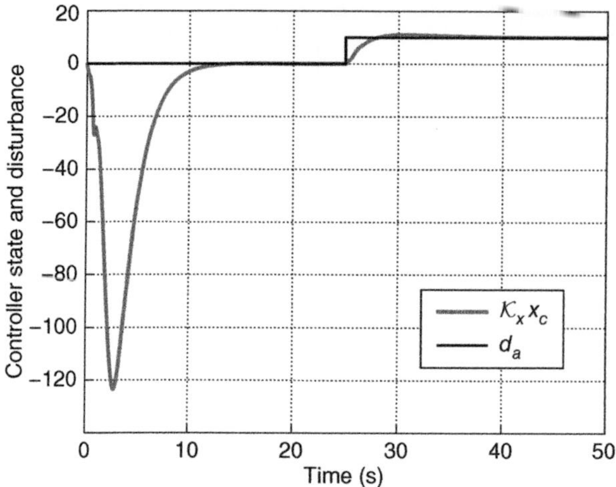

Figure 8.8 Time history of the matched disturbance d_a, and the controller state x_c multiplied by the constant $\mathcal{K}_x$.

https://youtu.be/JWqGukrjs44. The simulations and animations were performed under the same scenario, which was previously described in this section.

8.8.4 Disk on Disk System

This system consists of a nonactuated disk that rolls without slipping on another disk, which is actuated (see (Ryu et al., 2013) for details of the model). The coordinates of the system are $q = [q_1, q_2]$, where q_1 is the angle of the actuated disk and q_2 is the deviation angle of the nonactuated disk respect to the upright position. See the Figure 8.9. The mass matrix, the input matrix, and the potential energy of the system are as follows:

$$M = \begin{bmatrix} m_{11} & m_{12} \\ m_{21} & m_{22} \end{bmatrix}, \quad G = \begin{bmatrix} 1 \\ 0 \end{bmatrix}, \quad V = m_2 g(r_1 + r_2)\cos(q_2) \tag{8.78}$$

with $m_{11} = r_1^2(m_1 + m_2)$, $m_{12} = -m_2 r_1(r_1 + r_2)$, and $m_{22} = 2m_2(r_1 + r_2)^2$. The parameters m_i and r_i are the mass and radius of the disk i, respectively, for $i = 1, 2$. The constant g is the gravity. The control objective is to stabilize the desired equilibrium $q^\star = (0, 0)$, which is open-loop unstable. Using the classical IDA-PBC method, we obtain the controller:

$$\begin{aligned} u_{\text{IDA}}(q, p) = {} & \alpha_1 m_1 g(r_1 + r_2)\sin(q_2) - \alpha_2 \left(q_1 - \alpha_3 q_2\right) \\ & - \frac{k_v}{k_1 k_2 - k_3^2} \left(k_3 p_1 - k_2 p_2\right), \end{aligned} \tag{8.79}$$

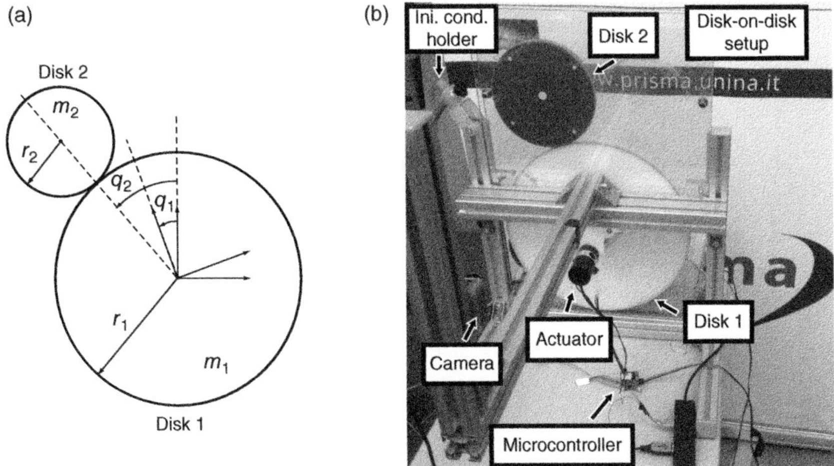

Figure 8.9 Idealized scheme of the Disk-on-Disk system (a) and real setup (b).

where

$$\alpha_1 := \frac{k_2 m_{11} - k_1 m_{12}}{k_3 m_{11} - k_2 m_{12}}, \quad \alpha_2 := k_4 \left(\frac{k_1 k_2 - k_2^2}{k_3 m_{11} - k_2 m_{12}} \right),$$

$$\alpha_3 := \frac{k_3 m_{12} - k_2 m_{22}}{k_2 m_{12} - k_3 m_{11}},$$

the damping injection gain $k_v > 0$, and the controller parameters k_1, k_2, k_3, and k_4 that satisfy

$$k_1 > 0, \quad k_1 k_3 - k_2^2 > 0, \quad k_4 > 0, \quad k_3 m_{11} - k_2 m_{12} > 0. \tag{8.80}$$

The desired mass matrix is

$$M_d = \begin{bmatrix} k_1 & k_2 \\ k_2 & k_3 \end{bmatrix} > 0,$$

and the desired potential energy is

$$V_d(q) = -\alpha_4 m_2 g(r_1 + r_2)\cos(q_2) + \frac{k_4}{2}(q_1 - \alpha_3 q_2)^2, \tag{8.81}$$

with $\alpha_4 = \frac{k_3 m_{11} - k_2 m_{12}}{m_{11} m_{22} - m_{12}^2}$. We first notice that the Disk-on-Disk verifies assumption **C12**. Thus, to reject the disturbances, we add to the IDA-PBC the outer-loop controller (8.57), (8.58) of Proposition 8.9 with the parameters K_1, K_3, K_I, and K_P to be chosen.

We carry out some experiments to evaluate the performance of the controller in the real setup shown in Figure 8.9. The model parameters are

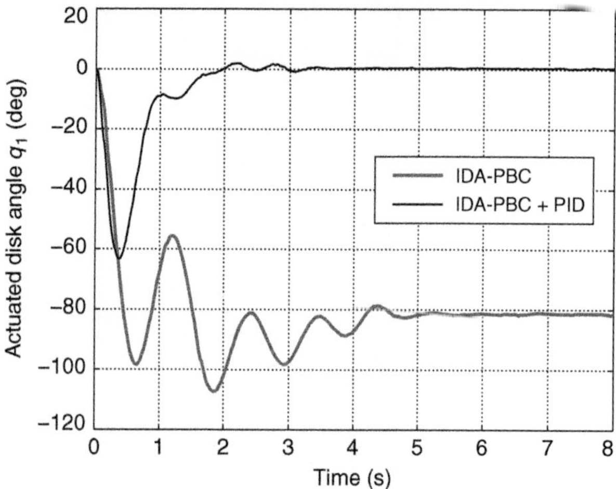

Figure 8.10 Time history of coordinate q_1 with the IDA-PBC controller plus the *nonlinear PID*. The initial condition is set to $q_1(0) = 0$ °.

$m_1 = 0.235$ kg, $m_2 = 0.0216$ kg, $r_1 = 0.15$ m and $r_2 = 0.075$ m. The parameters of the controllers were chosen as follows: $k_1 = 0.41$, $k_2 = -0.03$, $k_3 = 0.003$, $k_4 = 0.00025$, $k_v = 0.3$, $K_1 = 0.012$, $K_3 = 0.06$, $K_I = 2.2$, and $K_P = k_v$.

In the experiment, we add a matched constant disturbance to the system of value $d_a = 0.01$ N m. Figures 8.10 and 8.11 show the time history of the angle of the disk q_1 and the balancing angle q_2. As can be seen the angle of

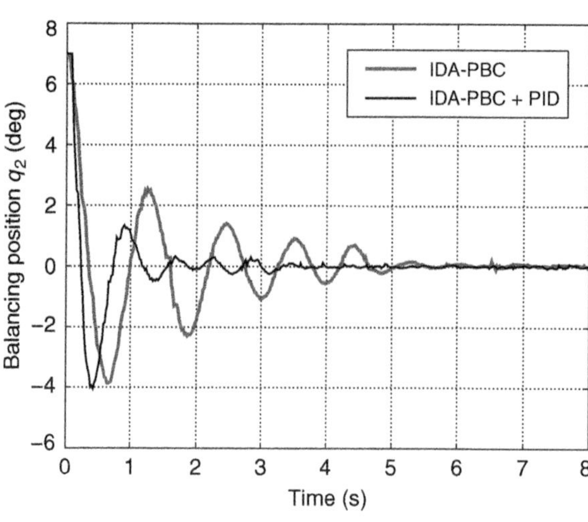

Figure 8.11 Time history of the coordinate q_2 with the IDA-PBC controller plus the *nonlinear PID*. The initial condition is set to $q_2(0) = 7$ °.

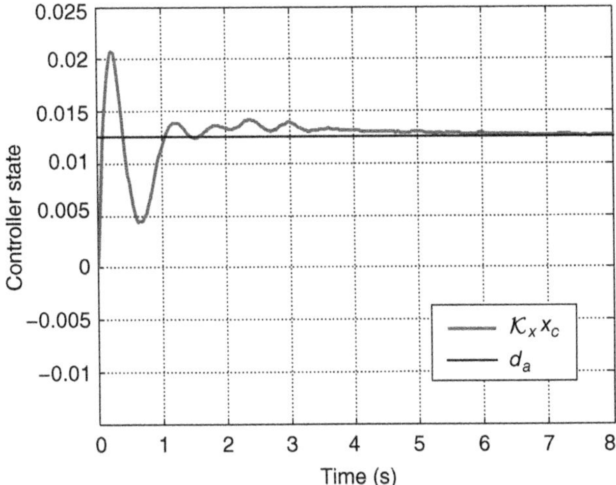

Figure 8.12 Time history of the matched disturbance d_a, and the controller state x_c multiplied by the constant $\mathcal{K}_x$.

the actuated disk q_1 reaches the desired value, and the nonactuated disk is balanced at the upright position, which is confirmed by the fact that q_2 converge to zero. The state of the controller x_c is shown in Figure 8.12 together with the disturbance, which shows that effectively the nonlinear PID compensates the action of the disturbance. Finally, the time history of the control input is depicted in Figure 8.13, which shows that the controller produces a physically reasonable torque.

A video of the experiments implementation of the disk-on-disk system in closed loop with IDA-PBC plus the nonlinear PID controller can be watched on https://youtu.be/JWqGukrjs44.

8.8.5 Damped Vertical Take-off and Landing Aircraft

The dynamics of the VTOL aircraft with uncertain physical damping in closed-loop with the IDA-PBC law proposed in Acosta et al. (2005), Gómez-Estern and van der Schaft (2004) are of the form (8.56) and friction matrix as (8.61), where $q = (x, y, \theta)$ with x, y denoting the translational position and θ the angular position of the aircraft, $\mathbf{p} = \dot{q}$ is the momentum,

$$V_d(q) = \frac{g(1-\cos\theta)}{k_1 - k_2\epsilon} + \frac{1}{2}[z(q) - z(q^\star)]^\top K_p [z(q) - z(q^\star)] \tag{8.82}$$

$$z(q) = \begin{bmatrix} x - x^\star - \frac{k_3}{k_1 - k_2\epsilon}\sin\theta \\ y - y^\star - \frac{k_3 - k_1\epsilon}{k_1 - k_2\epsilon}(\cos\theta - 1) \end{bmatrix} \tag{8.83}$$

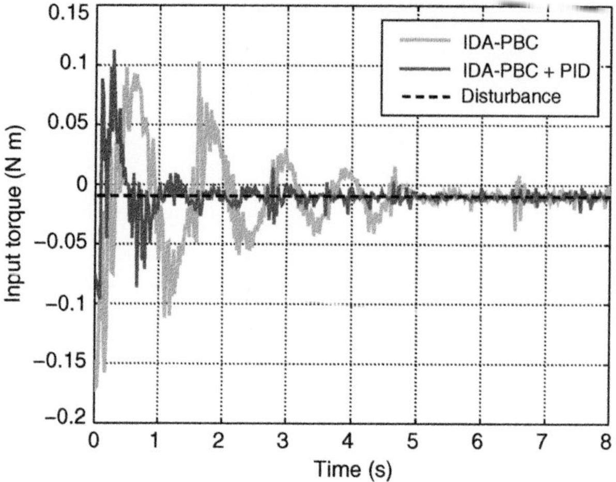

Figure 8.13 Time history of the control input and the disturbance with the sign changed.

$$
\begin{aligned}
\mathbf{M}_d(q) &= \begin{bmatrix} k_1\epsilon\cos^2\theta + k_3 & k_1\epsilon\cos\theta\sin\theta & k_1\cos\theta \\ k_1\epsilon\cos\theta\sin\theta & -k_1\epsilon\cos^2\theta + k_3 & k_1\sin\theta \\ k_1\cos\theta & k_1\sin\theta & k_2 \end{bmatrix} \\
D_p(q) &= \mathrm{diag}(r_1(q), r_2(q), r_3(q)) \\
G_u(q) &= \begin{bmatrix} 1 & 0 \\ 0 & 1 \\ \frac{1}{\epsilon}\cos\theta & \frac{1}{\epsilon}\sin\theta \end{bmatrix} \\
\mathbf{J}_2(q,\tilde{p}) &= \begin{bmatrix} 0 & \tilde{p}^\top\alpha_1(q) & \tilde{p}^\top\alpha_2 \\ -\tilde{p}^\top\alpha_1(q) & 0 & \tilde{p}^\top\alpha_3 \\ -\tilde{p}^\top\alpha_2 & -\tilde{p}^\top\alpha_3 & 0 \end{bmatrix} \\
\tilde{p}(q,\mathbf{p}) &= \mathbf{M}_d^{-1}(q)\mathbf{p} \\
\alpha_1(q) &= -\frac{1}{2}k_1\gamma_{30}\left[2\epsilon\cos\theta \quad 2\epsilon\sin\theta \quad 1\right]^\top \\
\alpha_2 &= -\frac{1}{2}k_1\gamma_{30}\left[0 \quad 1 \quad 0\right]^\top \\
\alpha_3 &= -\frac{1}{2}k_1\gamma_{30}\left[-1 \quad 0 \quad 0\right]^\top \\
\gamma_{30} &= k_1 - \epsilon k_2,
\end{aligned}
\tag{8.84}
$$

g is the acceleration due to gravity, ϵ is a constant that describes the coupling effect between the translational and rotational dynamics, $r_i(q) > 0$ are the damping coefficients, and $u \in \mathbb{R}^2$ is a control input for additional control design. The parameters $K_d, K_p \in \mathbb{R}^{2\times2}$, and $k_1, k_2, k_3 \in \mathbb{R}$ are tuning parameters of the IDA-PBC law that should be chosen to satisfy the conditions given in Section 7.1 of (Gómez-Estern and van der Schaft, 2004) to ensure stability of the closed-loop. Note that we have altered the dynamics from the cited sources with the addition of a constant matched disturbance for the purpose of illustration.

In the absence of external disturbance or any additional inputs, the equilibrium $(q, \mathbf{p}) = (q^\star, 0)$ is asymptotically stable (Gómez-Estern and van der Schaft, 2004). However, the presence of an external disturbance has the potential to perturb or destabilize the target equilibrium. Further complicating this situation, it is assumed that the open-loop damping matrix $D_p(q)$ is not exactly known and cannot be used for control purposes. Rather, an estimate $\hat{D}_p(q)$ is available and assumed to satisfy

$$\left\| D_p(q) - \hat{D}_p(q) \right\| \leq \kappa \tag{8.85}$$

for some constant $\kappa > 0$. In order to preserve the target equilibrium and stability properties, the integral action control (8.70) and (8.71) can be applied.

As the IA law (8.70), (8.71) is derivative of (8.35), assumption **C10** must be verified in order to guarantee asymptotic stability of the target equilibrium. As discussed in Section 8.7.3, the fact that the open-loop damping matrix D_p is full rank implies that the matrix U_u can be represented by (8.69). Furthermore, as all modeling error is assumed to be associated with the open-loop damping matrix, (8.34) is satisfied, implying that assumption **C10** is verified. Thus, the IA law (8.70), (8.71) can be applied directly to the VTOL system with uncertain damping by substituting in the values (8.82).

Bibliography

J. A. Acosta, R. Ortega, A. Astolfi, and A. D. Mahindrakar. Interconnection and damping assignment passivity-based control of mechanical systems with underactuation degree one. *IEEE Transactions on Automatic Control*, 50(12): 1936–1955, 2005.

A. Ailon and R. Ortega. An observer-based controller for robot manipulators with flexible joints. *Systems & Control Letters*, 21: 329–335, 1993.

D. Angeli, E. D. Sontag, and Y. Wang. A characterization of integral input-to-state stability. *IEEE Transactions on Automatic Control*, 45(6): 1082–1997, 2000.

T. Aoki, Y. Yamashita, and D. Tsubaniko. Vibration suppression for mass-spring-damper systems with a tuned mass damper using interconnection and damping assignment passivity-based control. *International Journal of Robust and Nonlinear Control*, 26(2): 235–251, 2016.

Z. Cai, W. Qu, Y. Xi, and Y. Wang. Stabilization of an underactuated bottom-heavy airship via interconnection and damping assignment. *International Journal of Robust and Nonlinear Control*, 17(18): 1690–1715, 2007.

A. Donaire and S. Junco. On the addition of integral action to port-controlled Hamiltonian systems. *Automatica*, 45(8): 1910–1916, 2009.

A. Donaire, J. Romero, R. Ortega, B. Siciliano, and M. Crespo. Robust IDA-PBC for underactuated mechanical systems subject to matched disturbances. *International Journal of Robust and Nonlinear Control*, 27(6): 1000–1016, 2017.

J. Ferguson, A. Donaire, and R. H. Middleton. Integral control of port-Hamiltonian systems: non-passive outputs without coordinate transformation. *IEEE Transactions on Automatic Control*, 62(11): 5947–5953, 2017a.

J. Ferguson, A. Donaire, R. Ortega, and R. H. Middleton. Matched disturbance rejection for energy-shaping controlled underactuated mechanical systems. In *IEEE Conference on Decision and Control (CDC)*, pages 1484–1489, Melbourne, Australia, 2017b.

J. Ferguson, A. Donaire, R. Ortega, and R. H. Middleton. Robust integral action of port-Hamiltonian systems. In *The IFAC Workshop on Lagrangian and Hamiltonian Methods for Nonlinear Control*, Valparaiso, Chile, 2018.

J. Ferguson, A. Donaire, R. Ortega, and R. H. Middleton. Matched disturbance rejection for a class of nonlinear systems. *IEEE Transactions on Automatic Control*, 65(4): 1710–1715, 2020.

J. Ferguson, R. H. Middleton, and A. Donaire. Disturbance rejection via control by interconnection of port-Hamiltonian systems. In *IEEE Conference on Decision and Control (CDC)*, pages 507–512, Osaka, Japan, 2015.

F. Gómez-Estern and A. J. van der Schaft. Physical damping in IDA-PBC controlled underactuated mechanical systems. *European Journal of Control*, 10(5): 451–468, 2004.

H. Khalil. *Nonlinear Systems*. Prentice-Hall, Upper Saddle River, NJ, 2002.

A. D. Mahindrakar, A. Astolfi, R. Ortega, and G. Viola. Further constructive results on interconnection and damping assignment control of mechanical systems: the acrobot example. *International Journal of Robust and Nonlinear Control*, 16(14): 671–685, 2006.

E. Nuño and R. Ortega. Achieving consensus of Euler-Lagrange agents with interconnecting delays and without velocity measurements via passivity-based control. *IEEE Transactions Control Systems Technology*, 26(1): 222–232, 2018.

R. Ortega and E. García-Canseco. Interconnection and damping assignment passivity-based control: a survey. *European Journal of Control*, 10(5): 432–450, 2004.

R. Ortega and J. G. Romero. Robust integral control of port-Hamiltonian systems: the case of non-passive outputs with unmatched disturbances. *System & Control Letters*, 61(1): 11–17, 2012.

R. Ortega, A. J. van der Schaft, I. Mareels, and B. M. Maschke. Putting energy back in control. *IEEE Control Systems Magazine*, 21(2): 18–33, 2001.

R. Ortega, M. W. Spong, F. Gómez-Estern, and G. Blankenstein. Stabilization of a class of underactuated mechanical systems via interconnection and damping assignment. *IEEE Transactions on Automatic Control*, 47(8): 1218–1233, 2002.

J. G. Romero, A. Donaire, and R. Ortega. Robust energy shaping control of mechanical systems. *System & Control Letters*, 62(9): 770–780, 2013.

J. C. Ryu, F. Ruggiero, and K. M. Lynch. Control of nonprehensile rolling manipulation: balancing a disk on a disk. *IEEE Transactions on Robotics*, 29(5): 1152–1161, 2013.

T. Shimizu, Y. Kobayashi, M. Sasaki, and T. Okada. Passivity-based control of a magnetically levitated flexible beam. *International Journal of Robust and Nonlinear Control*, 19(6): 662–675, 2009.

E. D. Sontag. *Input to State Stability: Basic Concepts and Results*, pages 163–220. Nonlinear and Optimal Control Theory. Springer, 2008.

A. J. van der Schaft. *L_2-Gain and Passivity Techniques in Nonlinear Control.* Springer-Verlag, Berlin, 3rd edition, 2016.

A. Venkatraman, R. Ortega, I. Sarras, and A. J. van der Schaft. Speed observation and position feedback stabilization of partially linearizable mechanical systems. *IEEE Transactions on Automatic Control*, 55(5): 1059–1074, 2010.

G. Viola, R. Ortega, R. Banavar, J. A. Acosta, and A. Astolfi. Total energy shaping control of mechanical systems: simplifying the matching equations via coordinate changes. *IEEE Transactions on Automatic Control*, 52(6): 1093–1099, 2007.

Appendix A

Passivity and Stability Theory for State-Space Systems

The following material is taken from van der Schaft (2016), Hill and Moylan (1980), and Khalil (2002). For further details, we refer the reader to these references.

A.1 Characterization of Passive Systems

The following definition characterizes *passive* systems.

Definition A.1 The state-space system Σ given in (1) is said to be *passive* if there exists a function $S : \mathbb{R}^n \to \mathbb{R}_+$, called the storage function, such that for all initial conditions $x(0) = x_0 \in \mathbb{R}^n$ the following inequality holds

$$S(x(t)) \leq S(x_0) + \int_0^t u^\top(s)y(s)ds. \tag{A.1}$$

Moreover, it is said to be

- *cyclo-dissipative* if $S : \mathbb{R}^n \to \mathbb{R}$ is not necessarily nonnegative;
- *lossless* if (A.1) holds with *equality*;
- *input strictly passive* if there exists $\delta > 0$ such that

$$S(x(t)) \leq S(x_0) + \int_0^t \left[u^\top(s)y(s) - \delta|u(s)|^2\right] dt.$$

- *output strictly passive* if there exists $\varepsilon > 0$ such that

$$S(x(t)) \leq S(x_0) + \int_0^t \left[u^\top(s)y(s) - \varepsilon|y(s)|^2\right] dt.$$

The following proposition establishes sufficient and necessary conditions to determine whether a system is passive, output strictly passive, or input strictly passive.

PID Passivity-Based Control of Nonlinear Systems with Applications, First Edition.
Romeo Ortega, José Guadalupe Romero, Pablo Borja, and Alejandro Donaire.

Published 2021 by John Wiley & Sons, Inc.

Proposition A.1: *Consider the system* Σ.

i) Σ *is passive* if and only if

$$\begin{bmatrix} 2[\nabla S(x)]^\top f(x) & [\nabla S(x)]^\top g(x) - h^\top(x) \\ g^\top(x)\nabla S(x) - h(x) & -\left[j(x) + j^\top(x)\right] \end{bmatrix} \leq 0, \ \forall x \in \mathbb{R}^n.$$

ii) Σ *is output strictly passive* if and only if

$$\begin{bmatrix} 2[\nabla S(x)]^\top f(x) + 2\varepsilon h^\top(x)h(x) & [\nabla S(x)]^\top g(x) - h^\top(x) + k^\top(x) \\ g^\top(x)\nabla S(x) - h(x) + k(x) & 2\varepsilon j^\top(x)j(x) - \left[j(x) + j^\top(x)\right] \end{bmatrix} \leq 0,$$

$$\forall x \in \mathbb{R}^n,$$

where

$$k(x) := 2\varepsilon j^\top(x)h(x).$$

iii) Σ *is input strictly passive* if and only if

$$\begin{bmatrix} 2[\nabla S(x)]^\top f(x) & [\nabla S(x)]^\top g(x) - h^\top(x) \\ g^\top(x)\nabla S(x) - h(x) & 2\delta I_m - \left[j(x) + j^\top(x)\right] \end{bmatrix} \leq 0, \ \forall x \in \mathbb{R}^n.$$

The Hill–Moylan theorem, given below, provides alternative – yet equivalent – conditions to determine if Σ is passive.

Theorem A.1 *Σ is passive with storage function $S : \mathbb{R}^n \to \mathbb{R}_+$ if and only if, for some $q \in \mathbb{N}$, there exist mappings $\ell : \mathbb{R}^n \to \mathbb{R}^q$ and $w : \mathbb{R}^n \to \mathbb{R}^{q\times m}$ such that*

$$\begin{aligned} [\nabla S(x)]^\top f(x) &= -\ell^\top(x)\ell(x), \\ h(x) &= g^\top(x)\nabla S(x) + 2w^\top(x)\ell(x), \\ j(x) + j^\top(x) &= w^\top(x)w(x). \end{aligned}$$

A.2 Passivity Theorem

To quote the passivity theorem the following definition is needed.

Definition A.2 The system Σ has $\mathcal{L}_2$-gain less than or equal to $\kappa > 0$ if

$$\|y\|_2^2 \leq \kappa\|u\|_2^2 + \beta, \ \beta \geq 0.$$

The following proposition relates the concepts of output strict passivity and $\mathcal{L}_2$-gain.

Proposition A.2: *If Σ is output strictly passive, then it has $\mathcal{L}_2$-gain less than or equal to $\frac{1}{\varepsilon}$.*

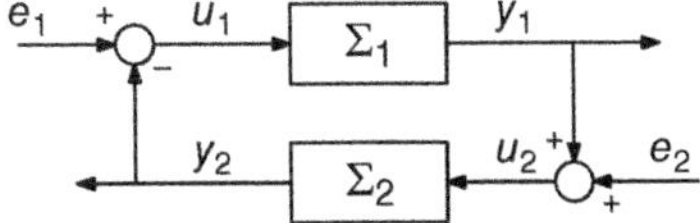

Figure A.1 Standard feedback configuration.

In this section we consider the classical negative feedback interconnection of two systems Σ_i, $i = 1, 2$, given in Figure A.1, assuming that the interconnection is well posed.[1] A first, obvious property is that if the subsystems Σ_i are passive, the interconnected system $\mathrm{col}(e_1, e_2) \to \mathrm{col}(y_1, y_2)$ is also passive with storage function $S_1(x_1) + S_2(x_2)$.

We are in the position to present a rather general version of the passivity theorem.

Proposition A.3: *Consider the feedback interconnection of Figure A.1, where each subsystem Σ_i verifies*

$$S_i(x_i(t)) \le S_i(x_i(0)) + \int_0^t \left[u_i^\top(s)y_i(s) - \varepsilon_i|y_i(s)|^2 - \delta_i|u_i(s)|^2\right] dt.$$

The interconnected system $\mathrm{col}(e_1, e_2) \to \mathrm{col}(y_1, y_2)$ *has finite $\mathcal{L}_2$-gain if*

$$\begin{aligned} \varepsilon_1 + \delta_2 &> 0, \\ \varepsilon_2 + \delta_1 &> 0. \end{aligned}$$

A.3 Lyapunov Stability of Passive Systems

The following definitions are necessary to present the content of this section.

Definition A.3 A function $S : \mathbb{R}^n \to \mathbb{R}$ is said to be $x^\star$-positive definite if

$$\begin{aligned} S^\star &= 0, \\ S(x) &> 0, \ \forall x \in \mathbb{R}^n \backslash \{x^\star\}. \end{aligned}$$

Definition A.4 Consider the system Σ with the output $y = h(x)$ satisfying $f(x^\star) = 0$. The output is said to be $x^\star$-*detectable* if

$$u(t) = 0, \ h(x(t)) = 0, \ \forall t \ge 0 \implies \lim_{t\to\infty} x(t) = x^\star.$$

The following theorem establishes a relation between passivity and stability, in the sense of Lyapunov, of an equilibrium point.

1 We adopt the natural notation $(\cdot)_i$ for states, inputs and outputs of the subsystems.

Theorem A.2 *Consider the system Σ with the output $y = h(x)$, satisfying $f(x^\star) = 0$. Assume the system is passive with a storage function $S(x)$, which is $x^\star$-positive definite and that the output y is $x^\star$-*detectable. *In this case, the feedback $u = -dy$, $d > 0$, asymptotically stabilizes the equilibrium $x^\star$. Furthermore, the equilibrium is globally asymptotically stable if $S(x)$ is radially unbounded.*

Bibliography

D. J. Hill and P. J. Moylan. Dissipative dynamical systems: basic input-output and state properties. *Journal of the Franklin Institute*, 309(5): 327–357, 1980.

H. Khalil. *Nonlinear Systems.* Prentice-Hall, Upper Saddle River, NJ, 2002.

A. J. van der Schaft. *L_2-Gain and Passivity Techniques in Nonlinear Control.* Springer-Verlag, Berlin, 3rd edition, 2016.

Appendix B

Two Stability Results and Assignable Equilibria

B.1 Two Stability Results

In this appendix two important stability results of nonlinear systems are revisited. We refer the reader to Krstić et al. (1995) and van der Schaft (2016) for the corresponding proofs.

Theorem B.1 *Consider the nonlinear time-varying system*

$$\dot{x} = f(x, t), \; x(0) = x_0, \tag{B.1}$$

where $x(t) \in \mathbb{R}^n$. *Let* $x^\star$ *be an equilibrium point for* (B.1) *and suppose* $f(x,t)$ *is locally Lipschitz in* x *uniformly in* t. *Let* $V : \mathbb{R}^n \times \mathbb{R}_+ \to \mathbb{R}_+$ *be a continuously differentiable function such that*

$$\begin{aligned} \gamma_1(\|x\|) \le V(x,t) \le \gamma_2(\|x\|), \\ \dot{V} \le -W(x) \le 0, \end{aligned} \qquad \forall t \ge 0, \; \forall x \in \mathbb{R}^n,$$

where γ_1, γ_2 *are class* $\mathcal{K}_\infty$ *functions, and* $W(x)$ *is a continuous function. Then, all solutions of* (B.1) *are* globally uniformly bounded *and satisfy*

$$\lim_{t \to \infty} W(x(t)) = 0.$$

Additionally, if $W(x)$ *is* positive definite and radially unbounded, *then* $x^\star$ *is* globally uniformly asymptotically stable.

Theorem B.2 *Consider the system* $\dot{x} = f(x)$ *where* $x(t) \in \mathbb{R}^n$ *and let* $\mathcal{M} \subset \mathbb{R}^n$ *be a compact set which is* positively invariant. *Let* $V : \mathbb{R}^n \to \mathbb{R}$ *be a differentiable function such that*

$$[\nabla V(x)]^\top f(x) \le 0, \; \forall x \in \mathcal{M}.$$

PID Passivity-Based Control of Nonlinear Systems with Applications, First Edition.
Romeo Ortega, José Guadalupe Romero, Pablo Borja, and Alejandro Donaire.

Published 2021 by John Wiley & Sons, Inc.

Every solution starting in $\mathcal{M}$ asymptotically converges to the largest invariant set in the set

$$\left\{x \in \mathbb{R}^n \mid [\nabla V(x)]^\top f(x) = 0\right\}.$$

B.2 Assignable Equilibria

Definition B.1 Consider the dynamical system Σ. We say that a vector $x^\star \in \mathbb{R}^n$ is an assignable equilibrium if there exists $u^\star \in \mathbb{R}^m$ such that

$$f(x^\star) + g(x^\star)u^\star = 0.$$

Proposition B.1: *The set of assignable equilibria Σ is given by*

$$\mathcal{E} = \left\{x \in \mathbb{R}^n \mid g^\perp(x)f(x) = 0\right\}, \tag{B.2}$$

where $g^\perp(x)$ denotes a full-rank-left annihilator of $g(x)$, i.e.,

$$g^\perp(x)g(x) = 0, \ \forall x \in \mathbb{R}^n. \tag{B.3}$$

Moreover, the following equivalence holds

$$f(x^\star) + g(x^\star)u^\star = 0 \iff \begin{bmatrix} g^\perp(x^\star)f(x^\star) = 0 \\ u^\star = -g^\dagger(x^\star)f(x^\star) \end{bmatrix}, \tag{B.4}$$

where $g^\dagger : \mathbb{R}^n \to \mathbb{R}^{m\times n}$ is the Moore–Penrose pseudoinverse of $g(x)$, i.e.,

$$g^\dagger(x) := \left[g^\top(x)g(x)\right]^{-1}g^\top(x). \tag{B.5}$$

Bibliography

M. Krstić, P. V. Kokotovic, and I. Kanellakopoulos. *Nonlinear and Adaptive Control Design*. John Wiley & Sons, Inc., New York, 1995.

A. J. van der Schaft. *L_2-Gain and Passivity Techniques in Nonlinear Control*. Springer-Verlag, Berlin, 3rd edition, 2016.

Appendix C

Some Differential Geometric Results

For completeness, in this appendix, we revisit some well-known differential geometric results that are used in the book. We refer the reader to Khalil (2002), Vidyasagar (1993), and van der Schaft (2016) for further details.

Throughout the appendix, we consider a dynamical system of the form

$$\dot{x} = f(x),\ x(0) = x_0, \tag{C.1}$$

where $x(t) \in \mathbb{R}^n$.

C.1 Invariant Manifolds

While the concept of *manifold* has a rigorous mathematical definition (Spivak, 1999), for our purposes, an $(n-m)$-*dimensional* manifold, with $1 \leq m < n$, is a set

$$\mathcal{M} := \{x \in \mathbb{R}^n \mid \gamma(x) = 0\}, \tag{C.2}$$

where $\gamma : \mathbb{R}^n \to \mathbb{R}^m$. The level sets

$$\mathcal{M}_\kappa := \{x \in \mathbb{R}^n \mid \gamma(x) = \kappa\}, \tag{C.3}$$

with $\kappa \in \mathbb{R}^m$, are called the *foliation* of the manifold $\mathcal{M}$.

Definition C.1 Let $\mathcal{A}$ be a subset of $\mathbb{R}^n$. Then, $\mathcal{A}$ is said to be a (positively) *invariant set* with respect to (C.1) if

$$x_0 \in \mathcal{A} \implies x(t) \in \mathcal{A},\ \forall t \geq 0.$$

Lemma C.1: *The* $(n-m)$-dimensional *manifold* $\mathcal{M}$ *given in* (C.2) *is invariant with respect to* (C.1) if and only if

$$[\nabla\gamma(x)]^\top f(x) = 0,\ \forall x(t) \in \mathcal{M}. \tag{C.4}$$

PID Passivity-Based Control of Nonlinear Systems with Applications, First Edition.
Romeo Ortega, José Guadalupe Romero, Pablo Borja, and Alejandro Donaire.

Published 2021 by John Wiley & Sons, Inc.

The foliation (C.3) is invariant with respect to (C.1) if and only if

$$[\nabla\gamma(x)]^\top f(x) = 0, \ \forall\, x(t) \in \mathbb{R}^n.$$

C.2 Gradient Vector Fields

Definition C.2 A *vector field* is a vector-valued function $\beta : \mathbb{R}^n \to \mathbb{R}^n$. It is said to be a *gradient* vector field if there exists a scalar function $\alpha : \mathbb{R}^n \to \mathbb{R}$ such that

$$\beta(x) = \nabla\alpha(x).$$

Lemma C.2: *$\beta : \mathbb{R}^n \to \mathbb{R}^n$ is a gradient vector field if and only if its Jacobian is symmetric, that is,*

$$\nabla\beta(x) = [\nabla\beta(x)]^\top.$$

C.3 A Technical Lemma

Definition C.3 Consider two vector fields $a : \mathbb{R}^n \to \mathbb{R}^n$, $b : \mathbb{R}^n \to \mathbb{R}^n$. Then, their *Lie bracket* is defined as

$$[a(x), b(x)] = \left[\nabla b(x)\right]^\top a(x) - [\nabla a(x)]^\top b(x). \tag{C.5}$$

Lemma C.3: *Consider a full rank map $G : \mathbb{R}^n \to \mathbb{R}^{n\times m}$. A necessary and sufficient condition for the existence of a coordinate transformation $z = \psi(x)$ such that*

$$[\nabla\psi(x)]^\top G(x) = \begin{bmatrix} I_m \\ 0 \end{bmatrix},$$

is that the columns of $G(x)$ commute. That is,

$$[G^i(x), G^j(x)] = 0, \ \forall i, j = 1, \dots, m,$$

where $G^i(x) \in \mathbb{R}^n$ are the columns of $G(x)$.

Bibliography

H. Khalil. *Nonlinear Systems*. Prentice-Hall, Upper Saddle River, NJ, 2002.

M. Spivak. *Comprehensive Introduction to Differential Geometry*. Perish, Inc., 3rd edition, 1999.

A. J. van der Schaft. *L_2-Gain and Passivity Techniques in Nonlinear Control*. Springer-Verlag, Berlin, 3rd edition, 2016.

M. Vidyasagar. *Nonlinear Systems Analysis*. 2nd edition. Prentice-Hall, Englewood Cliffs, NJ, 1993.

Appendix D

Port–Hamiltonian Systems

In this appendix, we define pH systems, prove their passivity properties, and discuss their relation with EL systems. See Duindam et al. (2009) and van der Schaft (2016) for further details.

D.1 Definition of Port-Hamiltonian Systems and Passivity Property

Definition D.1 The dynamics of a pH system are given as

$$\dot{x} = [\mathcal{J}(x) - \mathcal{R}(x)]\,\nabla H(x) + g(x)u \tag{D.1a}$$

$$y = \left[g(x) + 2\phi^{\top}(x)w(x)\right]^{\top}\nabla H(x) + \left[w^{\top}(x)w(x) + D(x)\right]u \tag{D.1b}$$

with state $x(t) \in \mathbb{R}^n$, input and output $u(t), y(t) \in \mathbb{R}^m$, respectively, $m \leq n$, Hamiltonian $H : \mathbb{R}^n \mapsto \mathbb{R}$, with $\mathcal{J}(x) = -\mathcal{J}^{\top}(x)$, $D(x) = -D^{\top}(x)$, $\mathcal{R}(x) = \mathcal{R}^{\top}(x) \geq 0$, and $\mathcal{R}(x) = \phi^{\top}(x)\phi(x)$.

Proposition D.1: *The input-state-output pH system (D.1) satisfies the power balance equation*

$$\dot{H} = -\begin{bmatrix}\nabla H^{\top}(x) & u^{\top}\end{bmatrix}\begin{bmatrix}\phi^{\top}(x)\phi(x) & \phi^{\top}(x)w(x)\\ w^{\top}(x)\phi(x) & w^{\top}(x)w(x)\end{bmatrix}\begin{bmatrix}\nabla H(x)\\ u\end{bmatrix} + u^{\top}y \leq u^{\top}y,$$

Hence, the map $u \mapsto y$ is cyclo-passive. It is passive if $H(x)$ is bounded from below.

PID Passivity-Based Control of Nonlinear Systems with Applications, First Edition.
Romeo Ortega, José Guadalupe Romero, Pablo Borja, and Alejandro Donaire.

Published 2021 by John Wiley & Sons, Inc.

D.2 Physical Examples

D.2.1 Mechanical Systems

The dynamics of simple mechanical systems can be written in pH form

$$\begin{bmatrix} \dot{q} \\ \dot{p} \end{bmatrix} = \begin{bmatrix} 0 & I_n \\ -I_n & 0 \end{bmatrix} \nabla H(q,p) + \begin{bmatrix} 0 \\ G(q) \end{bmatrix} \tau, \tag{D.2}$$

$$y_0 = G^\top(q) \nabla_p H(q,p), \tag{D.3}$$

where $q(t), p(t) \in \mathbb{R}^n$ are the generalized coordinate and momentum vectors, respectively, $\tau \in \mathbb{R}^m$, with $m \leq n$, are the generalized forces, $y_0 \in \mathbb{R}^m$ is the natural output, and $G(q) \in \mathbb{R}^{n \times m}$ is the input matrix. The Hamiltonian is

$$H(q,p) = \frac{1}{2} p^\top M^{-1}(q) p + V(q), \tag{D.4}$$

where $M(q) = M^\top(q) > 0$ is the inertia matrix, and $V(q)$ the potential energy function.

D.2.2 Electromechanical Systems

A multiport magnetic field electromechanical systems consisting of n_E magnetic ports and n_M mechanical ports is described by the pH model

$$\begin{bmatrix} \dot{\lambda} \\ \dot{q} \\ \dot{p} \end{bmatrix} = \begin{bmatrix} -R & 0 & 0 \\ 0 & 0 & I_{n_M} \\ 0 & -I_{n_M} & -R_M \end{bmatrix} \nabla H(\lambda, q, p) + \begin{bmatrix} Bu \\ 0 \\ F_L \end{bmatrix}, \tag{D.5}$$

where $\lambda \in \mathbb{R}^{n_E}$ is the vector of flux linkages, $q \in \mathbb{R}^{n_M}$ the (rotational or translational) generalized coordinates and $p \in \mathbb{R}^{n_M}$ are the momenta of the masses. The system's total energy function is

$$H(\lambda, q, p) := H_E(\lambda, q) + H_M(q, p),$$

where $H_E(\lambda, q)$ is the magnetic energy stored in the inductances and $H_M(q, p)$ is the mechanical energy of the form (D.4). $R \geq 0$ is the resistor matrix in series with the inductances, and $R_M \geq 0$ accounts for the Coulomb friction effects, B is a constant, full rank, input matrix to the electrical ports, and we have included external voltages u and mechanical forces F_L. This model contains, as particular cases, electric motors/generators and magnetic levitation systems.

For the sake of completeness, we recall that the constitutive relations of the elements are

$$\begin{aligned} i &= \nabla_\lambda H_E(\lambda, q), \\ F &= -\nabla_q H_E(\lambda, q), \\ \dot{q} &= \nabla_p H_M(q, p), \end{aligned}$$

where i are the inductor currents, F are the mechanical forces of electrical origin, and the minus sign of the second equation reflects Newton's third law.

D.2.3 Power Converters

The average dynamics of a large class of power converters – operating with sufficiently fast sampling rate – is described by the pH model

$$\dot{x} = \left(J_0 + \sum_{i=1}^{m} J_i u_i - R \right) \nabla H(x) + \left(G_0 + \sum_{i=1}^{m} G_i u_i \right) E, \tag{D.6}$$

where $x(t) \in \mathbb{R}^n$ is the converter state, $u(t) \in \mathbb{R}^m$ denotes the duty ratio of the switches, $J_i = -J_i^\top,\ i \in \bar{m} := \{0, \dots, m\}$ are the interconnection matrices, $R = R^\top \geq 0$ represents the dissipation matrix, $G_i,\ i \in \bar{m}$, are $n \times n$ input matrices, and the vector $E \in \mathbb{R}^n$ contains the external voltage and current sources, which may be switching. Assuming linear elements, the total energy stored in inductors and capacitors is

$$H(x) = \frac{1}{2} x^\top Q x,$$

with $Q = Q^\top > 0$ determined by the values of the capacitances and inductances. See Escobar et al. (1999) for further details on these models.

D.3 Euler–Lagrange Models

As shown in Ortega et al. (1998) the dynamics of many physical systems – including mechanical systems – can be described in the EL form

$$\frac{d}{dt} \left[\nabla_{\dot{q}} \mathcal{L}(q, \dot{q}) \right] - \nabla_q \mathcal{L}(q, \dot{q}) = G(q) u, \tag{D.7}$$

where $\mathcal{L} : \mathbb{R}^{n \times n} \mapsto \mathbb{R}$ is the Lagrangian function

$$\mathcal{L}(q, \dot{q}) = \frac{1}{2} \dot{q}^\top M(q) \dot{q} - V(q), \tag{D.8}$$

$q \in \mathbb{R}^n$ are generalized coordinates and $u \in \mathbb{R}^m$ generalized forces acting on the system. Using (D.8) in (D.7), we obtain the classical EL model given as

$$M(q)\ddot{q} + C(q, \dot{q})\dot{q} + \nabla V(q) = g(q) u, \tag{D.9}$$

where the following key *skew-symmetry* property holds

$$\dot{M}(q) = C(q, \dot{q}) + C^\top(q, \dot{q}). \tag{D.10}$$

In the case of mechanical systems, $C(q, \dot{q})$ is the Coriolis matrix.

The relationship between the EL and the pH representation is established via the Legendre transform of $\mathcal{L}(q, \dot{q})$, implicitly defined as

$$H(q, p) := \dot{q}^\top p - \mathcal{L}(q, \dot{q}),$$
$$p := \nabla_{\dot{q}} \mathcal{L}(q, \dot{q}). \tag{D.11}$$

See (van der Schaft, 2016, Section 4.5) for further details.

D.4 Port-Hamiltonian Representation of GAS Systems

The following result, called "Universal Stabilization" property of interconnection and damping assignment-passivity-based control (IDA-PBC) in Ortega et al. (2002), see also Wu et al. (2020), establishes the connection between GAS systems and their pH representation.

Proposition D.2: *Assume the equilibrium $x^\star$ of the system $\dot{x} = f(x)$ is GAS, with $f(x)$ Lipschitz continuous. Then, there exist a C^1 positive definite function $H(x)$ and C^0 matrices $\mathcal{J}(x) = -\mathcal{J}^\top(x)$ and $\mathcal{R}(x) = \mathcal{R}^\top(x) \geq 0$ such that*

$$f(x) = [\mathcal{J}(x) - \mathcal{R}(x)]\nabla H(x).$$

Bibliography

V. Duindam, A. Macchelli, S. Stramigioli, and H. Bruyninckx. *Modeling and Control of Complex Physical Systems: The Port-Hamiltonian Approach.* Springer Science & Business Media, 2009.

G. Escobar, A. J. van der Schaft, and R. Ortega. A Hamiltonian viewpoint in the modeling of switching power converters. *Automatica*, 35(3): 445–452, 1999.

R. Ortega, J. A. Loría, P. J. Nicklasson, and H. Sira-Ramírez. *Passivity-Based Control of Euler-Lagrange Systems.* Springer-Verlag, 1998.

R. Ortega, A. J. van der Schaft, B. M. Maschke, and G. Escobar. Interconnection and damping assignment passivity–based control of port–controlled Hamiltonian systems. *Automatica*, 38(4): 585–596, 2002.

A. J. van der Schaft. *L_2-Gain and Passivity Techniques in Nonlinear Control.* Springer-Verlag, Berlin, 3rd edition, 2016.

D. Wu, R. Ortega, and G. Duan. On universal stabilization property of interconnection and damping assignment control. *Automatica*, 119, 109087 2020.

Index

PID Passivity-Based Control of Nonlinear Systems with Applications, First Edition.
Romeo Ortega, José Guadalupe Romero, Pablo Borja, and Alejandro Donaire.

Published 2021 by John Wiley & Sons, Inc.

Printed and bound by CPI Group (UK) Ltd, Croydon, CR0 4YY